BIOLOGY for TODAY

Book 1

Ernest G. Neal

MBE, M.Sc., Ph.D., F.I.Biol.

formerly Second Master and Head of the Science Department
Taunton School

and

Keith R. C. Neal

M.A., F.I.Biol.

Head of the Biology Department
Manchester Grammar School

Line illustrations by Barry Jones and Marion Mills

Blandford Press   Poole   Dorset

First published in the U.K. 1974 by
Blandford Press Ltd,
Link House, West Street,
Poole, Dorset BH15 1LL

Reprinted (with corrections) 1976
Reprinted 1978
Reprinted 1980

ISBN 0 7137 0658 9 (School Edition)
ISBN 0 7137 0674 0 (Library Edition)

Filmset by Keyspools Ltd, Golborne, Lancs
Printed in Great Britain by
Butler and Tanner Ltd, Frome, Somerset

Authors' Preface

In recent years the publication of enquiry-based courses in Biology has greatly stimulated the teaching of the subject, and these have already led to a welcome change of emphasis in the traditional courses. We recognise the great importance of the enquiry approach and believe that many topics are admirably suited to this method; however, we feel that other topics are better explained in a more traditional manner. Therefore we have not hesitated to incorporate both these approaches in the course. It is also our belief that a text-book at this level should provide a clear framework of easily accessible factual information and that the text, illustrations and practical work should be closely integrated in order to lead the student towards a clear understanding of biological principles.

The course is designed to meet the requirements of the syllabuses of the various Examining Boards for the General Certificate of Education at Ordinary Level in Biology.

We are convinced that any O-level course in Biology should be educational in the widest sense of the word and should provide future citizens with a better understanding of themselves and their environment. For this reason considerable emphasis has been given to Man and to the application of biological principles to human affairs. We have also given special prominence to subjects such as behaviour, ecology and conservation because of their great relevance to life in the 1980's.

The arrangement of topics in a Biology course is always a matter of individual preference. We have divided this course roughly into three sections. In Book 1 the first 11 chapters introduce the student to basic biological principles through the study of a wide variety of animals and plants. The remaining chapters are primarily functional in approach and deal in greater detail with Man and flowering plants. This is continued in the first 9 chapters of Book 2. The final section is concerned with wider topics including Man and his environment, genetics and evolution, and draws together many of the principles from previous chapters.

Throughout the course we have kept in mind the changing interests and powers of comprehension of the student which are associated with increasing age and experience, but the order can easily be adjusted to suit the individual preferences of the teacher.

We have tried to include adequate instructions in the text for the practical work to be carried out successfully, but some additional information, together with suggestions for relevant teaching aids, has been included in the Appendix.

Nobody can write a text-book without being greatly indebted to many people. We are very conscious of the great debt we owe to the authors of the many works we have consulted and would specially mention those concerned with the Nuffield Science Teaching Project and the Scottish Course, Biology by Inquiry. Some of the experiments and ideas in this book have been based on these and other courses. But an author's gratitude can only partly be covered by formal acknowledgments because equally important are the many helpful discussions and influences which can less easily be attributed to an individual.

Particular thanks must go to Mr John Haller of Philip Harris Biological Ltd., who has given most practical help in the provision of photographs, many of them taken by himself, and for his enthusiastic cooperation over the whole project. The line illustrations have been specially drawn by two artists, Barry Jones and Marion Mills. We are very grateful to them for their close cooperation over each diagram and for producing such outstanding work. A detailed list of acknowledgments relating to the illustrations will be found on p. 208. It is regretted if any credit has been unwittingly omitted.

We are most grateful to the following who have read sections of the manuscript and given us the benefit of their advice, criticism and encouragement: Dr David Bygott, Mr Howard Green, Miss Muriel Hosking, Mr Martin Jacoby, Mr Andrew Neal, Dr Philip Penny, Mr Gordon Perry, the late Mr Colin Russell, Mr Geoffrey Stephens and Dr David Watson; also Mrs Hazel Watson and Mrs Elizabeth Wells for typing the manuscript. Finally, we warmly thank our wives for reading the manuscript critically, for their many valuable suggestions and for their patience and encouragement.

E.G.N.   K.R.C.N.

Contents

1

Introduction to biology

What is biology?

Biology is the study of life—the study of all living things. Because we are alive, it also includes the study of ourselves. We share this amazing characteristic of being alive with bacteria and fungi, ferns and worms, buttercups and frogs, fish and spiders, horses and lions.

One striking fact about living organisms is their fantastic variety. They range from the microscopic plants and animals found in a drop of pond water to the giant redwood trees of California and the majestic blue whales of the Antarctic ocean. Some form of life is found practically everywhere on earth, even in unlikely places such as hot springs and deserts. But no matter how deeply we study animals and plants, life itself still retains an element of mystery and grandeur which, perhaps, we will never fully understand.

Why study biology?

Biology brings a new appreciation of life. It helps to open our eyes to the amazing variety and marvellous complexity of living things. It stirs our imagination and increases our sense of wonder. Look down a microscope at a drop of pond water and a new world comes into view; a world of strange, exciting organisms, each a marvel of intricate construction. A swallow flying overhead may appear to some as just another bird, but think of its perfect command of flight in all weathers and the navigational skill which enables it to migrate many thousands of kilometres and return the

next year to the exact place where it had previously nested. Watch a bee busily collecting pollen from a flower. Is it just another insect? An insect, yes, but one that is capable of steering by sun-compass to track down a patch of flowers and then pass on to other bees the exact location of that source of nectar.

We often marvel at our own technology, and rightly so, but it is a humbling thought that many of man's achievements are often clumsy imitations of phenomena found in living things. For example, millions of years before man discovered radar, bats were using a very similar technique for locating objects in the dark. Man still has much to learn from nature. Biology is a young science and research prospects for the further benefit of mankind are limitless. There is still much to discover and much to wonder at.

Those of you who hope one day to become doctors or nurses, veterinary surgeons or farmers, horticulturalists or foresters, pharma-

Fig. 1:1 Biology includes the study of ourselves.

Fig. 1:2 Gannet colony on Grassholm, S. Wales.

cists or bacteriologists, naturalists or conservationists, to give just a few examples, will find that your work is concerned with the application of biological principles. But whatever job we do in life, biology is most important because it helps us to understand ourselves. Studying the structure of animals and the way they live helps us to understand how our *own* bodies work; observing the behaviour of animals and how they react with each other, and with their environment, helps us to understand why *we* behave as we do.

Man, like other animals, is basically dependent on plants for food; his survival depends on the success of his crops. Every day we use products made from substances acquired from plants and animals. Many of our clothes are made from cotton or wool and the leather for our shoes originates from the hides of animals; our newspapers are made from wood pulp and the tyres of cars and bicycles are manufactured from the latex of rubber trees. We need a home in which to live and bring up a family. Think of your own home and see how much of it originated from plants—wood is one of our basic raw materials. Although synthetic substances are replacing some of these products it is clear that man cannot do without animals and plants. They are essential for his existence.

Without an understanding of how plants and animals live, man is in great danger of over-exploiting them. Already irreparable damage has been done and man has only just begun to realise that subduing nature is not enough; in order to survive he must live in harmony with it. This is not going to be easy as his rapid rise in numbers and the increase in his standard of living are putting more and more strain on available resources. To succeed, biological understanding will be essential. It is an exciting challenge to all of us.

What have all living organisms in common?

Although there is so much diversity among living organisms there are certain characteristic features which they all have in common, and this, of course, applies to us too. They can be summarised as follows:

1. Nutrition. All plants and animals need **food** or the raw materials for making it. The ways in which a grass, a mushroom and a cow obtain it are all different, but the function of the food is the same: to provide energy for all living processes and to provide the raw materials necessary for growth, for the replacement of worn-out parts of the body and for the reproduction of the species. Nutrition is the term used to describe both the taking in of food (feeding) and the subsequent chemical changes that take place before it is used. Green plants feed differently from all other living things as they alone can use light energy to build up sugar from carbon dioxide and water and to build up other complex foods with the help of simple salts. All other organisms have to use food already made by plants or other animals and have to digest it first before it can be absorbed into the body.

Fig. 1:3 Flowers in abundance; an alpine meadow.

2

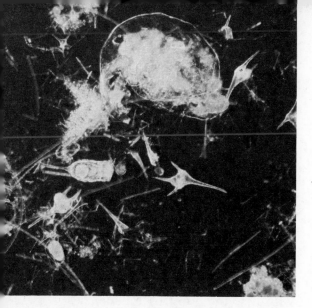

Fig. 1:4 Some of the forms of life found in a drop of lake water.

2. Respiration. This is the process whereby energy is released from food. For this to happen **oxygen** is usually required and **carbon dioxide** and **water** are given off as by-products. The methods of obtaining oxygen vary a great deal; for example when a fish pumps water over its gills the oxygen in the water diffuses into the blood which circulates through them, but in a jellyfish or a tree no special movements are apparent. The important point is that respiration, the release of energy, takes place in every cell of the body—in both plants and animals.

3. Excretion. The chemical processes, such as respiration, taking place within the cells and tissues of an organism are all described by the term **metabolism**. One result of these processes is that waste substances are produced. If they built up to too high a level they would be harmful, so through the process of excretion these substances are removed.

Excretion can therefore be defined as the removal of the waste products of metabolism. It differs from **defaecation**, the removal of waste matter from the alimentary canal, in that most of this waste matter has not originated from processes taking place within the cells of the body. Substances in the urine, however, are truly excretory as they have been formed in the body and removed from the blood by the kidneys.

4. Growth. This is the permanent increase in size of an organism due to the formation of new living matter or **protoplasm**. For this to take place there must be an adequate supply of food. Except for the simplest organisms, growth also leads to an increase in complexity. Thus the germination and growth of an acorn lead eventually to a very complex oak tree. It is incredible, too, to think that each one of us has grown from just one fertilized egg cell about one-fifth of a millimetre in diameter!

5. Reproduction. This is the process whereby life itself is passed on from one generation to another. Living organisms cannot exist indefinitely—the bodies of animals and plants eventually die through old age, disease, accident or the inability to cope with environmental conditions. Reproduction usually takes place when the organism has reached its maximum size, although you will be able to think of exceptions. There are special mechanisms for ensuring that the characteristics of the parents are passed on to the offspring. Thus the eggs of a robin always hatch into robins rather than any other kind of bird.

6. Irritability. In order for organisms to survive (and thus to reproduce) they must be able to respond to changes in their environment, a characteristic known as irritability (or sensitivity). This involves being able to detect a particular **stimulus**, such as light, and to make an appropriate **response** to it. You may have noticed that when you expose a plant to one-sided illumination its shoot responds very slowly by bending towards the light; this is a growth response. But if you shine a light on to an earthworm half out of its burrow at night, it rapidly disappears; the response in this case is one of movement. Movement is such a characteristic response to a stimulus, especially in animals, that some biologists prefer to consider it separately as a seventh characteristic of all living things, but we consider it more logical to retain it as an example of irritability.

The examples given above of responses to stimuli are advantageous to the organism. If a plant shoot did not grow towards the light it might eventually die, because light is essential for the manufacture of its food. Similarly, if an earthworm did not quickly disappear down its burrow when dawn came it would soon be eaten by a bird. You might think what other stimuli animals and plants respond to and whether these responses help them to survive. Consider also your own responses to stimuli.

3

Living and non-living things

Some inanimate objects appear to show some of the characteristics of living things. A computer will 'respond' when certain 'stimuli' are fed into it; a crystal of copper sulphate will 'grow' when kept in a saturated solution of the salt. A motor car is 'fed' with petrol, it 'respires' it with the help of oxygen (combustion), the energy released is used to propel the car, and the waste products are 'excreted' through the exhaust system. However, the fundamental difference between living and non-living things is that a living organism controls all the processes itself. For example, in the case of the car, somebody has to fill the petrol tank, turn on the engine and drive it along. Even the incredibly complex and seemingly automatic movements of space craft are ultimately controlled by the men who write the computer programs.

Viruses—living or non-living?

It is not easy to say whether viruses are living or non-living. Some can be extracted, crystallised and stored in a bottle indefinitely. They are minute particles, so small that they can only be seen with the electron microscope. They will reproduce, but only when they are inside a cell of a living organism; it seems that they take over the chemical machinery of the cell and use it to make more of themselves. These peculiar structures which only show some of the characteristics of living things will be considered further in Chapter 19.

Branches of biology

Because the science of biology covers such a wide field, various terms are used to describe particular areas of investigation.

Taxonomy is concerned with classification, that is, the sorting out of animals and plants into groups according to their common characteristics. The term **morphology** is often used to describe the general form of an organism, but internal morphology is more commonly known as **anatomy**. Human anatomy, for example, deals with the structure and arrangement of all the parts of the body such as bones, blood vessels, nerves and muscles. In contrast, **physiology** is concerned with how these organs function, e.g. how a muscle contracts. **Biochemistry** is really a branch of physiology but concentrates more on the chemistry of these processes. The study of the early development of a plant or animal is known as **embryology**. In man this would cover the period from the fertilization of the egg to birth. The science of animal behaviour is known as **ethology**, whereas the study of the relationships between animals and plants and their environment comprises **ecology**. **Evolution** is the branch of biology which attempts to explain how new kinds of organisms may have originated. This is closely linked with the study of heredity, or **genetics**, which is concerned with the way the characteristics of organisms are passed on from one generation to another.

2

Cells and organisms

How are plants and animals constructed?

In 1665, **Robert Hooke** described how he cut some very thin slices of cork and examined them under his extremely simple microscope. He found that its structure resembled a honeycomb, so he called the units of which it was composed **cells**. He also looked at thin sections of plant stems and roots and found that they also consisted of cells, though not always of the same shape. Later it was discovered that the bodies of animals were also made of cells.

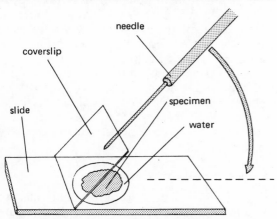

Fig. 2:1 Method of lowering a coverslip on to a preparation.

Our bodies are made of billions of them; a single drop of blood contains about 5 million! Cells can be described as the units of living matter, the microscopic 'bricks' from which animals and plants are made.

Examine various kinds of cells for yourself.

1. Onion cells. Cut an onion into four quarters. It consists largely of white scale leaves which easily separate. It is the thin skin or **epidermis** covering the inner surface of one of these scale leaves that you should examine. You will need two pieces about 5 mm square. Prepare the first piece by cutting the epidermis with a sharp scalpel; then peel it off carefully with forceps and mount it in a drop of water on a slide. Prod it gently to remove any air bubbles which may cling to it and place a coverslip on top (Fig. 2:1). Repeat for the second square, but mount it in iodine solution instead of water.

Examine the preparation in water first. Have all the cells the same number of sides? Do the walls appear to be rigid? Can you see under the high power any sign of the living protoplasm? This consists of two distinct parts, a rounded structure in the cell, the **nucleus**, and some almost transparent material which lines the cell wall, the **cytoplasm**. The latter is not easily seen. Now examine your iodine preparation. Does it show up better? Biologists use many kinds of stains to make things clearer. Draw 2 or 3 cells carefully.

2. Moss cells. Mount the leaf of a moss in a drop of water and examine its cells. Do you notice an important difference from the cells of the onion? The large number of rounded green bodies which make the whole leaf look green are called **chloroplasts**. They are typical of most plant cells which are exposed to light. Make a drawing of a few of these cells.

3. Cheek cells. As an example of animal cells, examine some of your own. Cells lining the inside of the cheek come off readily when rubbed gently. First sterilise the *handle* of a scalpel. If it is made of metal, pass it through a flame several times and wait until it is cool. If it is made of wood, dip it in alcohol and allow

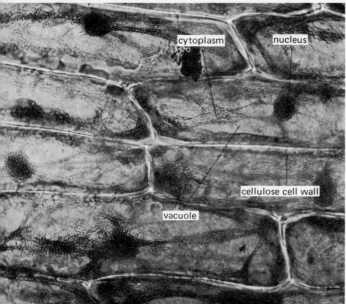

a

Fig. 2:2 High power photomicrographs of various cells:
a) from the epidermis of a scale leaf of an onion bulb
b) from the lining of the cheek c) from a moss leaf.

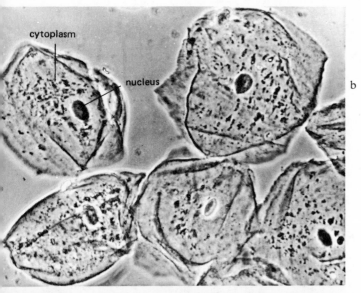b

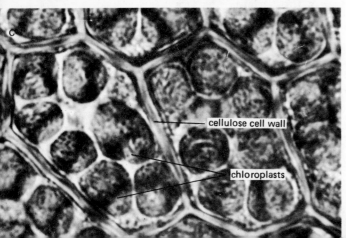c

the latter to evaporate. Now use it to rub the inside of your cheek. Put a small drop of saliva on a slide and stir the scalpel handle in it to get off some of the cells. Mount in iodine as before and examine under a microscope. Some cells will be separate, others will still be attached in groups. Draw some of them under the high power and compare them with the plant cells.

4. Examine specially prepared slides of thin sections of parts of plants and animals.

You will now realise that cells of plants and animals are of many shapes and sizes. Being alive, they all have protoplasm; thus they all normally have two things in common, the possession of a nucleus and cytoplasm, the two forms in which protoplasm occurs. The nucleus controls the activities of the cell and plays an important part when cells reproduce, while the cytoplasm carries out all the complicated processes going on in the cell.

Although the nucleus and cytoplasm looked very simple under your microscope, a photograph taken under an electron microscope (Fig. 2:3) reveals that protoplasm is extremely complex with various parts carrying out different functions.

How do plant and animal cells differ?

Typical plant cells have rigid cell walls made of a non-living organic substance called **cellulose**, while animal cells merely have a very thin, flexible membrane, the **plasma membrane**, which is part of the living cytoplasm. Plant cells usually have spaces or **vacuoles** in them which are filled with a watery fluid called **cell sap**; if they are present in animal cells they are usually so small that they cannot easily be seen under an ordinary microscope. Animal cells contain no chloroplasts while the majority of plant cells do; but those deep inside stems or roots have none as light is necessary for their formation.

Why are cells not all alike?

When a house is built, different materials are used in certain places to serve different purposes; likewise, the cells of the body of an organism differ according to their function.

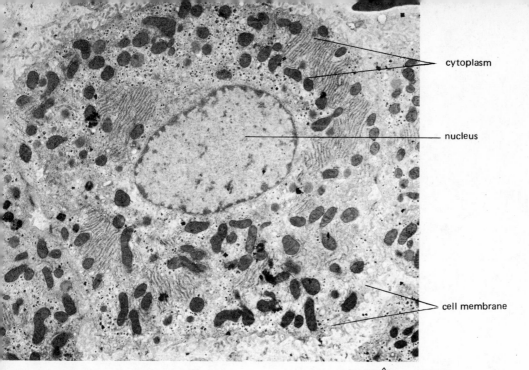

Fig. 2:3 Electronmicrograph of a liver cell.

Fig. 2:4 Animal cells. 1 Involuntary muscle. 2 Sperm. 3 White blood cell. 4 Ciliated cell from lining of wind pipe. 5 Nerve cell from brain.

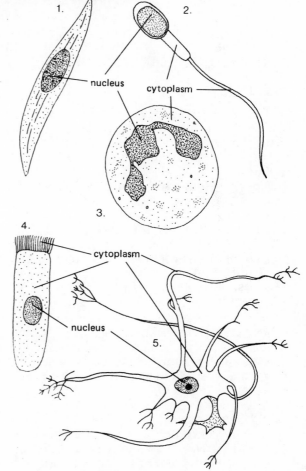

Thus in our own body, **muscle** cells are able to contract and relax and are used in movement; **nerve** cells are able to receive and send out nervous impulses through their long processes; cells from the lining of the stomach secrete digestive juices on to the food, and **sperms** are motile cells which have a long tail by which they can swim towards an egg and fertilize it (Fig. 2:4). Similarly, in plants there are **mesophyll** cells in the leaf which use their chloroplasts to build up food, **parenchyma** cells which act as packing between the veins and help to keep a stalk rigid, and **phloem** cells which transmit food from one part of the plant to another (Fig. 2:5). All are adapted in their structure to fulfil a particular function.

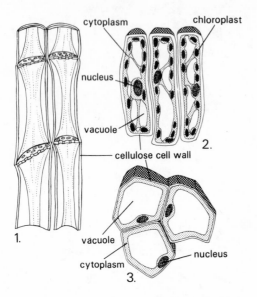

Fig. 2:5 Plant cells. 1 Phloem conducting cells from a stem. 2 Mesophyll cells of a leaf. 3 Parenchyma cells.

Fig. 2:6 (above) Epithelial cells from the lining of the small intestine; (below) cartilage cells.

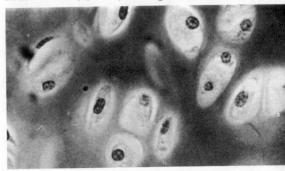

HOW OUR BODY IS BUILT UP

Tissues

Cells are not arranged in a haphazard way, but usually occur in groups according to their type. These aggregates of similar cells which carry out a particular function are called **tissues**.

There are four main groups of tissues making up our bodies:

1. Epithelia (sing. epithelium)
These are sheets of cells which line the inside, or cover the outside of the various parts of the body, e.g. the lining cells of the stomach and the surface layer of the skin.

2. Connective tissues
These are very variable. Some help to bind parts of the body together, others like bone and cartilage help to support the body, others such as the blood are fluid and are used for transport.

3. Muscular tissues
These are concerned with movement. Some are used for strong contractions as in limbs, others for rhythmic contractions such as those in the walls of the gut which help squeeze the food along, others in the heart wall keep up regular movements causing the blood to be pumped round the body.

4. Nervous tissues
These are concerned mainly with the conduction of nervous impulses and are concentrated in the brain and spinal cord.

Organs

These are more complex structures which are built up from various tissues and perform a specific function. Thus the heart is the organ which pumps the blood, the stomach is the organ which digests the food, the kidneys are the organs which excrete urine, and the lungs are the organs which are concerned with breathing.

Systems

Organs are often grouped together to form systems which serve a general purpose. The main systems of our body are:
1. The **alimentary system** which is concerned with the intake, digestion and absorption of food and the elimination of undigested material.
2. The **respiratory system** which brings oxygen into the body and gives out the unwanted carbon dioxide and water.

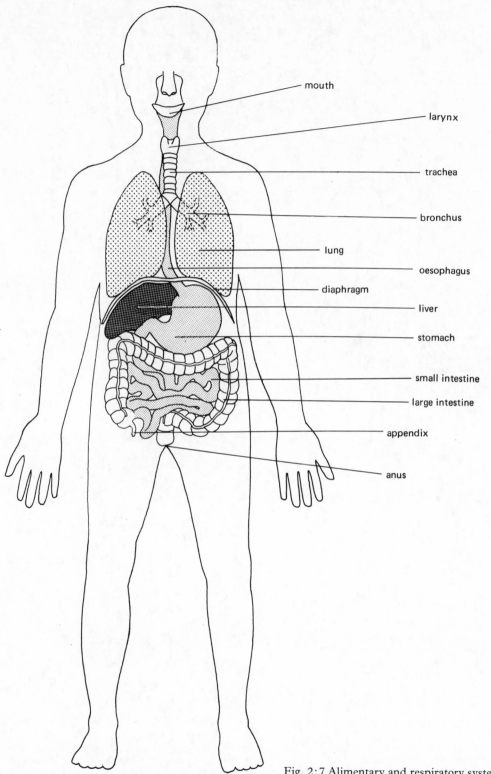

mouth

larynx

trachea

bronchus

lung

oesophagus

diaphragm

liver

stomach

small intestine

large intestine

appendix

anus

Fig. 2:7 Alimentary and respiratory systems of man.

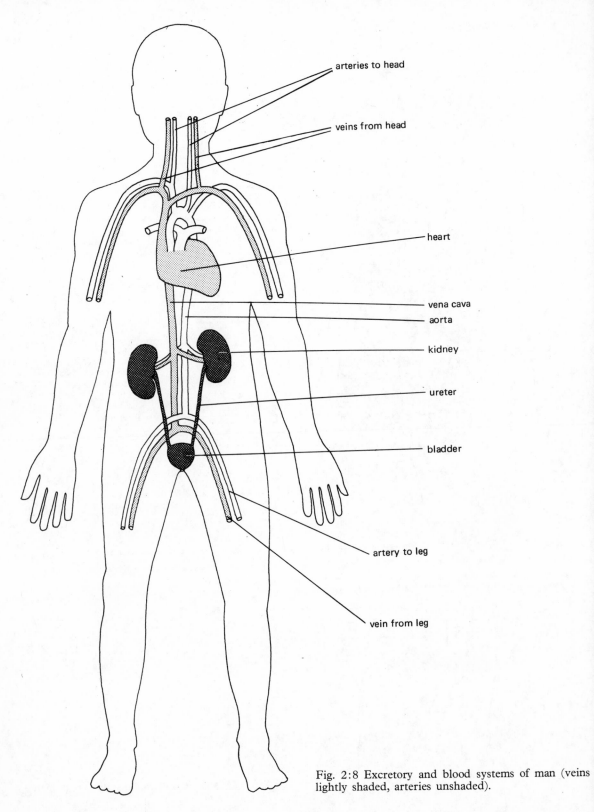

arteries to head

veins from head

heart

vena cava

aorta

kidney

ureter

bladder

artery to leg

vein from leg

Fig. 2:8 Excretory and blood systems of man (veins lightly shaded, arteries unshaded).

10

3. The **reproductive system** which produces the sex cells (sperms in male, eggs in female) and which protects and nourishes the young during development.

4. The **blood system** which consists of heart and blood vessels and enables materials to be transported to all the other organs and tissues.

5. The **skeletal system** which comprises the bones which give support and protection to other parts of the body.

6. The **muscular system** for movement of the whole body or its parts.

7. The **excretory system** which eliminates useless or harmful products which have been made in the body.

8. The **nervous system** which controls and co-ordinates the actions of the body and consists of the brain, spinal cord and nerves.

9. The **sensory system** which receives information from outside and inside the body and consists of sense organs such as the eyes, ears and nose.

These systems do not work independently; they all co-operate with one another so that the body functions as a whole—a complete **organism**. You will see from Figs. 2:7 and 2:8 how some of these organs and systems are arranged; also watch an animal such as a rat being dissected or look at a prepared dissection to see the marvellous way in which all the organs and systems fit together. In Chapter 5 we shall see how plants are also divided into systems.

DIVISION OF LABOUR

This is an important principle which arises from what we have learnt in this chapter about cells and their functions. Division of labour means the specialisation of different structures for various functions. We have seen that this occurs at many levels:

1. Division of labour within the cell, e.g. the nucleus for cell division and control, the chloroplasts for the manufacture of food.

2. Division of labour between cells or groups of similar cells (tissues), e.g. muscles for movement, bones for support.

3. Division of labour between organs, e.g. the

Fig. 2:9 Levels of organisation within a mammal.

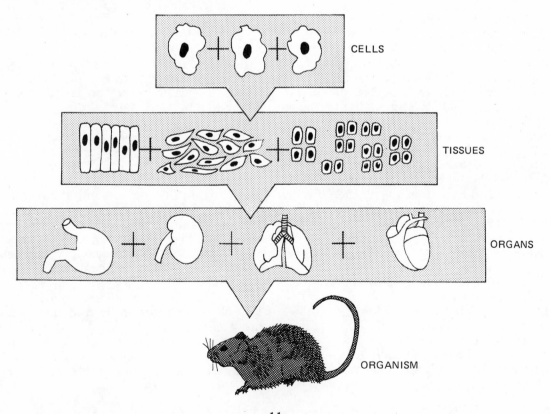

CELLS

TISSUES

ORGANS

ORGANISM

11

heart for pumping blood, the stomach for digestion.

There is also a fourth level of division of labour between individual organisms living in the same colony or society, e.g. in a bee colony where there are queens, drones and workers which are specialised to serve the society in different ways (p. 77).

We have also seen how the structure of cells and organs is always related to the functions they perform. Later on we shall see an extension of this principle in that the structure of whole organisms is also closely related to the way they live and the place they live in.

SIMPLE ORGANISMS

We are now going to study some of the simplest organisms of all, those which have no tissues or organs, but nevertheless can carry out all the properties of living things. The two we shall choose are *Amoeba*, one of the simplest animals alive today, and *Spirogyra* which is a simple plant.

AMOEBA

Amoebae are just visible to the naked eye as tiny white specks. Most species live in freshwater where they move about on the mud at the bottom. Some live in the sea and others in damp soil. A few species are parasitic and may cause diseases such as amoebic dysentery. You will probably examine a rather large species *Amoeba proteus* which is readily cultured in the laboratory (Fig. 2:10).

Amoebae belong to the group of animals known as Protozoa (p. 20). These animals are not divided into cells and are best described as **non-cellular**, although in many ways they resemble a single cell and are often called **unicellular** in consequence.

Structure

Watch an amoeba under the high power of a microscope or look at a film of one. Notice how this tiny grey blob of protoplasm constantly changes its shape as it flows along. Note how the cytoplasm is clear and jelly-like near the surface (**ectoplasm**) and granular and more watery inside (**endoplasm**). Some of the granules in the endoplasm act as food reserves,

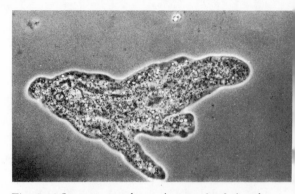

Fig. 2:10 Low power photomicrograph of *Amoeba*.

others are excretory. There is no cell wall, and you may wonder why the contents of the amoeba do not just dissolve in the water and the animal does not disintegrate. It does not do so because of its **plasma membrane**, an extremely thin 'skin' forming the outer layer of the ectoplasm. This holds it together and controls to some extent which substances go in and out.

Look for the nucleus. It cannot always be seen in a living specimen but you see it best when the amoeba is actively moving. It usually appears as a slightly oblong, rather transparent structure being carried along in the moving cytoplasm, but if viewed end-on, it is seen to be biconcave.

Look also for a transparent spherical object, the **contractile vacuole**. This controls the amount of water inside the animal and acts rather like a safety valve. You may have to focus carefully to see it. An amoeba cannot avoid absorbing water all the time through its surface by a process called osmosis (p. 121); it collects in the contractile vacuole which gradually enlarges until it suddenly disappears, squeezing out its contents into the surrounding water. It then starts enlarging again. You may be fortunate enough to see this happening, although it is a slow process in amoeba; in some species of Protozoa, e.g. *Paramecium*, it occurs much more frequently.

Amoeba appears to be very simple in structure, although under an electron microscope it has been shown to be very much more complex. The amazing thing is that this minute mass of protoplasm is a complete organism capable of carrying out all the properties of life.

12

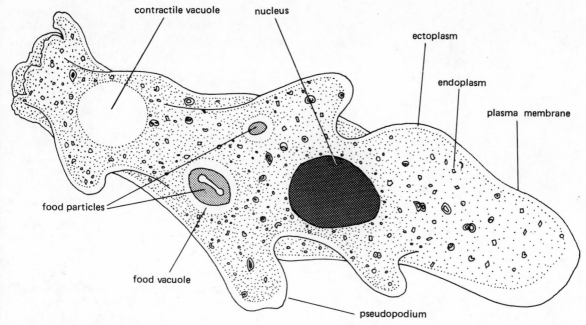

contractile vacuole   nucleus

ectoplasm

endoplasm

plasma membrane

food particles

food vacuole

pseudopodium

Fig. 2:11 *Amoeba proteus.*

Irritability

Directional movement is a response to an external stimulus. Although it has no eyes, the cytoplasm is sensitive to light; an inactive amoeba when illuminated soon starts to move. It is also sensitive to touch, and may move away from a solid object which it meets. It can also detect chemical substances in the water; harmful substances cause it to move away or to become spherical, while others which emanate from food cause it to move towards them.

When it moves, a small bump of ectoplasm is first formed, then the fluid endoplasm seems to flow into it to form a protuberance called a **pseudopodium** which enlarges as more endoplasm flows into it from other parts. Sometimes several pseudopodia are put out at the same time while the cytoplasm of others is being withdrawn.

Nutrition

Amoeba feeds on bacteria, protozoa or algae much smaller than itself. As a result of contact, pseudopodia are put out all round the food to enclose it in a drop of water, now called a **food vacuole** (Fig. 2:12). The cytoplasm round the vacuole pours digestive juices into it and most of the food is made soluble and absorbed into the cytoplasm. Any undigested material is removed when the vacuole comes to the surface and breaks.

Respiration and excretion

Amoeba, like any other organism, needs energy and in the process of respiration this is liberated from the food by means of oxygen. The oxygen diffuses through its surface from the surrounding water. Carbon dioxide is formed which diffuses out through the surface with other excretory substances in solution.

Growth and reproduction

With good feeding an amoeba increases in size rapidly, but when it reaches a certain size the nucleus divides into two (Fig. 2:13) and the cytoplasm begins to constrict as the two nuclei move apart; eventually the two parts separate as new individuals. This method of asexual (non-sexual) reproduction is called **binary fission** and is one of the simplest

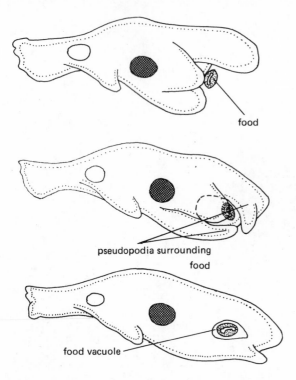

food

pseudopodia surrounding
food

food vacuole

Fig. 2:12 Method of feeding in *Amoeba*.

Fig. 2:13 Asexual reproduction in *Amoeba*.

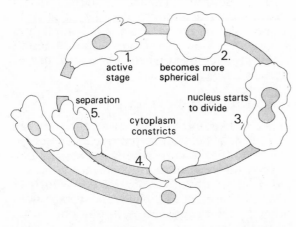

1. active stage
2. becomes more spherical
separation 5.
nucleus starts to divide
cytoplasm constricts 3.
4.

methods. No sexual method (one resulting from the fusion of cells) is known in amoeba.

Some large amoebae are also capable of producing minute spores within themselves, each spore containing some nuclear material and part of the cytoplasm. The amoeba eventually breaks up and liberates these spores which have resistant coats. If the pond dries up these spores can remain dormant for long periods, and with the return of damp conditions they

hatch out and grow into amoebae. In the dormant dry state they may be blown to other places and start new colonies. Large amoebae can also form hard protective cysts round themselves if conditions are bad and then they remain dormant for a period.

SPIROGYRA

In contrast to *Amoeba*, which is a non-cellular animal, *Spirogyra* is a simple green plant. You find this species in ponds and ditches as a mass of hair-like, dark green filaments which are slimy to touch because of a film of mucilage on their surface.

Structure

Examine some of the filaments in a drop of water under the low power. Notice how each plant is a single cylindrical filament consisting of many similar cells arranged end to end. Under the high power, notice the rigid cell wall made of cellulose and the conspicuous ribbon-like chloroplasts which wind spirally along each cell. Different species have different numbers of chloroplasts in each cell; the one in Fig. 2:14 has two.

You will see other structures, and if you add a drop of iodine they will show up better.

Do you notice any change of colour caused by the iodine? Iodine stains most things brown, but it turns anything containing starch, blue-black. Can you see any spherical structures at intervals along the chloroplast which have turned almost black? They are called **pyrenoids**, bodies which store starch.

Now look for the nucleus in the centre of each cell; it stains dark brown in contrast to the lighter brown of the cytoplasm which surrounds it. Look also for the cytoplasm which lines the cell wall and the threads which connect it to the cytoplasm around the nucleus. The spaces contained by the cytoplasm form the vacuole which is filled with cell sap.

Nutrition

Spirogyra, like other green plants, builds up sugar by a process called **photosynthesis** from water and carbon dioxide which it ab-

14

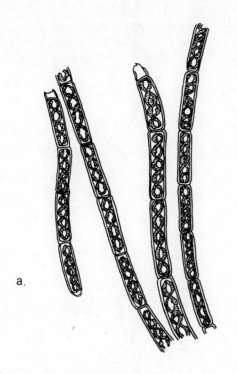

a.

sorbs from the surrounding water. Both light energy and the chlorophyll in the chloroplast are needed for the process and the sugar is then turned to starch for storage in the pyrenoids. *Spirogyra* also obtains salts from the water which are used in the formation of other complex substances such as proteins.

Respiration and excretion

Energy is released from the sugar formed in photosynthesis by respiration with the help of oxygen which is dissolved in the surrounding water and absorbed through the cell wall. Carbon dioxide formed as a by-product is excreted in the reverse direction.

Growth

Enlargement of the filament takes place when the nucleus of each cell divides into two and a new cell wall is formed between them (Fig. 2:15). The cells then gradually increase in size until they get as big as the original cells, when they divide again. Thus the filament grows longer.

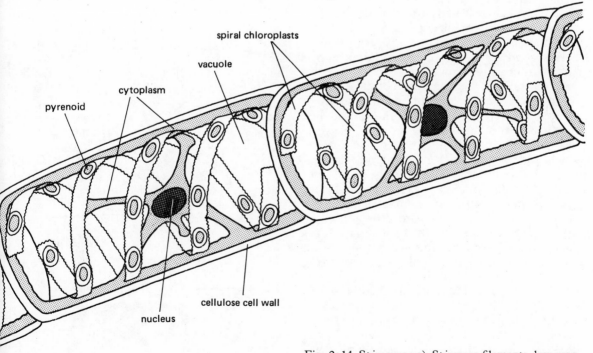

spiral chloroplasts

vacuole

cytoplasm

pyrenoid

nucleus

cellulose cell wall

b.

Fig. 2:14 *Spirogyra:* a) *Spirogyra* filaments, low magnification b) two cells enlarged.

15

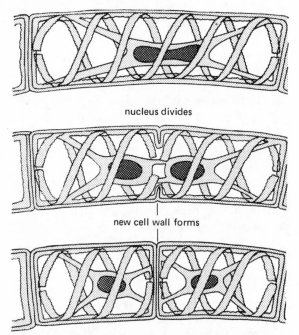

nucleus divides

new cell wall forms

Fig. 2:15 A filament of *Spirogyra* grows when each of its cells divides and the new cells then enlarge.

Reproduction

This occurs when more filaments are formed. One method is called **fragmentation** when a filament simply breaks up into smaller parts; this is an asexual process.

A second method is more complex and is known as **conjugation** (Fig. 2:16). It is a sexual process involving the fusion of the contents of two cells and results in the formation of **zygospores** with hard resistant walls which protect them if the water dries up. *Spirogyra* has two kinds of filaments which look exactly alike, but behave differently during conjugation. Normally reproduction only occurs between two filaments of different kinds. If two such filaments come into contact lengthwise, swellings form from opposite cells and these push the strands apart as they get larger. (This is an example of irritability—a response to a contact stimulus.) Eventually the separating wall breaks down to form a canal which connects the two cells. Meanwhile the cytoplasm of each cell becomes a rounded mass, the chloroplasts break up and the contents of one cell pass down the tube and fuse with that of the opposite partner, nucleus fusing with nucleus. This fusion is called **fertilization**.

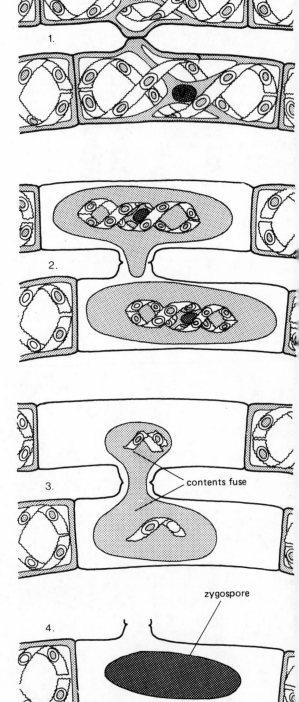

1.

2.

3.

contents fuse

4.

zygospore

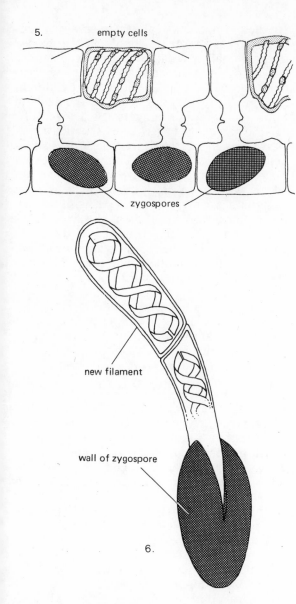

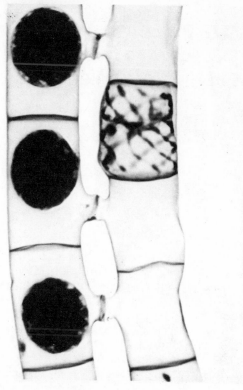

Fig. 2:17 Photomicrograph of *Spirogyra* showing zygospores.

Fig. 2:16 Conjugation of *Spirogyra*: 1. Swellings occur. 2. Separating wall breaks down. 3. Contents of one cell passes through conjugation canal. 4. A zygospore is formed following the fusion of the contents of both cells. 5. Part of two filaments after conjugation. 6. Germination of a zygospore.

This process may take place all down the filament to give a ladder-like appearance, the cells of one plant being empty while those of the other contain zygospores. When the filament eventually decays, the zygospores may remain dormant during the winter in the mud, be carried to other parts by the water current, or be dispersed to other watery habitats on the feet of birds.

Thus this simple plant, consisting of cells which are all alike, carries out all the functions characteristic of living things. When we compare *Spirogyra* with *Amoeba*, we notice differences in the ways they both carry out these functions and these illustrate the differences between plants and animals generally.

3

Classifying organisms

The classification of organisms is complex because there are over a million different kinds of animals and about 343 thousand kinds of plants which are known. The problem is sorting them out, according to their similarities, into well-defined groups and giving them each a different name.

What similarities are the best ones to use in such a classification? If you use such features as colour or size, method of movement or number of legs, you find you have grouped together a very odd assortment of animals and produced an extremely **artificial classification**. This is not very helpful. The classification we use today is based on structural similarities. This has the great advantage in that it groups together organisms which are nearly related. It is therefore described as a **natural classification**.

The modern theory is based on the work of the great Swedish naturalist, **Linnaeus**, who as long ago as 1758 described his method of classification in a book called *Systema Naturae* (The Order of Nature). One of his problems was to think of a different name for every kind of plant and animal. Obviously there were not enough names to go round! This is what he did: just as we have at least two names, a surname which we share with our brothers and sisters, and a first name which distinguishes us from them, so Linnaeus gave each kind of organism two names. This method was therefore called the **binomial classification**. The name corresponding to the surname is the name of the **genus** which it shares with nearly-related forms, the other is the name of the **species**. Linnaeus realised it was very important that the name should be recognised internationally. What confusion would arise if a harmful pest or a disease-producing organism was called by different names in different countries! He therefore named them all in Latin as this was a universal language. The generic name is put first. Thus he called a man *Homo sapiens*, a frog *Rana temporaria* and a pine tree *Pinus sylvestris*. You will notice that the generic name is usually a noun and always begins with a capital letter and the specific name is descriptive and is spelt with a small letter. For example, different species of ladybird have a characteristic number of spots; the common one is called *Coccinella septempunctata* (meaning 7-spotted), another *C. bipunctata* (2-spotted) and a third *C. decempunctata* (10-spotted).

By using two names the specific name may be used many times for different organisms as for each organism the generic name will be different; thus *Pieris brassicae* is the large white butterfly and *Barathra brassicae* is the cabbage moth. *Brassicae* means 'of the cabbage', and is given to both because the caterpillars of both species feed on cabbages.

Linnaeus used seven main categories in his classification. The largest was the **kingdom** of which he recognised two, plants and animals. Today with our knowledge of bacteria and some of the other microscopic organisms it is not always easy to say whether some of them are plants or animals, so it is quite usual to group some of these simple forms into a third kingdom, the **Protista**. However, for our purpose we will keep to the simpler classification and use two kingdoms only.

Each kingdom is sub-divided into large groups called **phyla** (sing. phylum), which in turn are divided into **classes** and these into **orders** which are composed of **families** which contain various **genera** (sing. genus) which include certain **species**.

Some examples are given in the table on the following page.

Let us consider these groups in more detail to see the sort of structural characteristics that are used in classification. If you refer to the table you will see that man and elephant are both put into the phylum Vertebrata; this is because their general plan of body structure

	Man	African elephant	Common ladybird	Scots pine tree	Beech tree
Kingdom	Animalia	Animalia	Animalia	Plantae	Plantae
Phylum	Vertebrata	Vertebrata	Arthropoda	Spermatophyta	Spermatophyta
Class	Mammalia	Mammalia	Insecta	Gymnospermae	Angiospermae
Order	Primates	Proboscidea	Coleoptera	Coniferales	Fagales
Family	Hominidae	Elephantidae	Coccinellidae	Pinaceae	Fagaceae
Genus	*Homo*	*Loxodonta*	*Coccinella*	*Pinus*	*Fagus*
Species	*H. sapiens*	*L. africana*	*C. septempunctata*	*P. sylvestris*	*F. sylvatica*

is similar, including a backbone composed of vertebrae. They are both placed in the class Mammalia because they show the characteristic of having mammary glands by which they suckle their young. But they are put in different orders because, for one reason, man has quite different teeth from an elephant and eats very different things; he also has a relatively larger brain. In this he is much more like the apes and monkeys, so he is put in the same order as these—the Primates. But there are obvious differences between man and the apes or the monkeys, so he is separated off into the family Hominidae which only includes man-like creatures. Other primitive types of man once belonged to this family, but they are now all extinct and there is only one species left which includes all the races of mankind living today. Look at the table below. Each line represents a different level of classification. Notice how there are great differences

between all the animals in the top line, rather fewer differences between those in the next, fewer still in the next, and so on until with animals belonging to the same genus the differences are extremely small.

Now let us take a plant example. In the above table you see that the Scots pine and the beech tree are both put in the phylum Spermatophyta; this is because they both produce seeds, but they are put into different classes because the pine has no flowers and forms seeds from ovules which are freely exposed, while the beech has flowers and forms seeds inside a fruit. The pine is put in the order Coniferales because the seeds are produced in special cones, and in the family Pinaceae because it shows, with other species of pine, the characteristic needle leaves which occur in small clusters.

Thus in every classification you start with large similarities and end with small ones.

Kingdom	Otter	Starfish	Snail	Insect	Worm	Jellyfish	
Phylum	Otter	Bird	Reptile	Amphibian	Fish		
Class	Otter	Sheep	Rabbit	Bat	Monkey		similarities
Order	Otter	Dog	Lion	Bear	Seal		becoming
Family	Otter	Badger	Stoat	Weasel			greater
Genus	European otter (*Lutra lutra*)	Canadian otter (*Lutra canadensis*)					↓

What is a species?

Members of a species are different in structure from all other organisms and they vary little between themselves apart from differences between male and female and variations due to living under different conditions. They can also interbreed and produce offspring which are themselves able to reproduce. Thus the lion and the tiger are separate species although they occasionally breed to produce tigrons, but the latter are normally incapable of having young themselves. The same applies to horses and donkeys when they produce mules. All dogs are considered to be the same species as theoretically they are all able to breed together and the offspring are fertile. The great variation in appearance between different breeds is due to man's selection of the puppies in a litter from which he wishes to breed.

From this definition of a species you will see why the races of mankind are all one species. The variation between the races, for example in skin colour, is due to the fact that they have lived under different conditions for many thousands of years.

Besides structural differences between species there are other factors of importance. Members of the same species often exhibit very similar behaviour; a song thrush in Scotland makes the same kind of nest and lines it with mud as does a song thrush in England; a grey squirrel buries nuts in the ground in a particular way whether it lives in America or in Britain. But although there are many *general* behavioural similarities, it has to be remembered that individual members of a species often behave very differently; consider the members of your class! Members of the same species are also very similar biochemically, i.e. in the sort of chemical reactions going on in the body and in the kind of substances produced; this applies particularly to the proteins which are formed. Thus similarities of structure, biochemistry, behaviour and the ability to produce fertile offspring are all characteristic of a species.

Now let us consider the classification of the animal kingdom as a whole. Here is a much simplified classification including only the most important phyla and classes; the illustrations will give you an idea of what they look like. The animals in the first eight phyla are often called invertebrates because they have no backbone.

The main animal phyla

1. PROTOZOA

Microscopic animals which are not made up of cells, e.g. *Amoeba, Paramecium, Euglena* and many parasites some of which cause diseases such as malaria and sleeping sickness.

2. COELENTERATA

Animals with a jelly-like body composed of two layers of cells only, with a central mouth surrounded by tentacles. Sometimes there is a hard skeleton on the outside as in corals. Other examples are jellyfish, anemones and *Hydra*.

3. PLATYHELMINTHES (Flatworms)

Worm-like creatures with flattened or ribbon-shaped bodies. Some called planarians, live in freshwater under stones, others are parasites in other animals, e.g. tapeworms and flukes.

4. NEMATODA (Roundworms)

Worms with round bodies, pointed at both ends, and no rings or segments. Many are parasites such as the hookworm, others are free-living in the soil.

5. ANNELIDA (Segmented worms)

Worms with round bodies marked externally into rings or segments. They may have one or more pairs of bristles on each segment. The main groups comprise the earthworms, marine worms and leeches (no bristles).

Fig. 3:1 Various invertebrates:
a) Coelenterata: 1. Portuguese man-o-war $\times \frac{1}{6}$. 2. Coral $\times 2$. 3. Hydra $\times 4$.
b) Platyhelminthes: 4. Tapeworm $\times 2$. 5, 6. Planarians. 7. Parasitic flukes $\times 4$.
c) Nematoda: 8. Parasitic worms $\times 20$.
d) Annelida: 9. Ragworm $\times 1$. 10. Leech $\times 1$. 11. Earthworm $\times 1$.
e) Mollusca: 12. Pond snail (coiled shell) $\times 2$. 13. Scallop (bivalve) $\times \frac{1}{2}$. 14. Squid $\times \frac{1}{2}$.
f) Echinodermata: 15. Sea urchin $\times \frac{1}{4}$. 16. Starfish $\times \frac{1}{3}$.

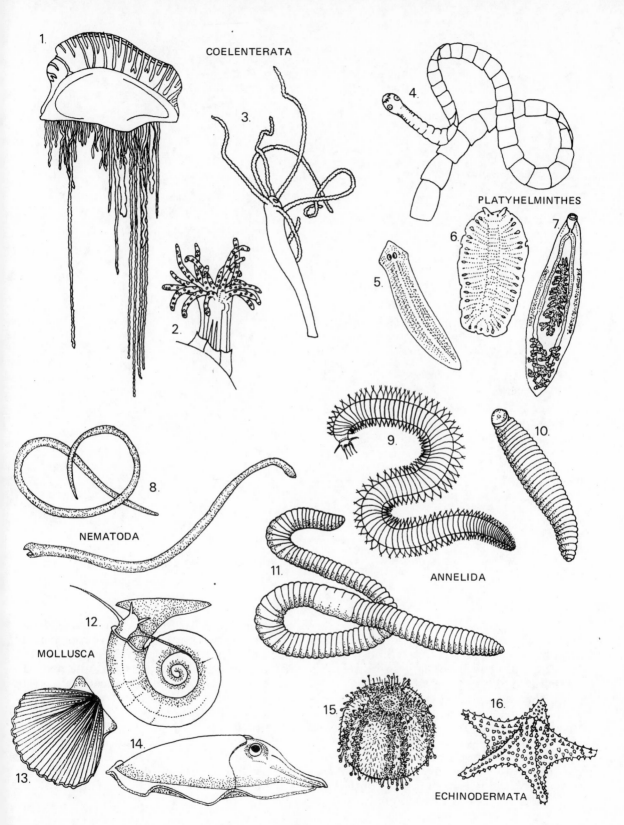

COELENTERATA

PLATYHELMINTHES

NEMATODA

ANNELIDA

MOLLUSCA

ECHINODERMATA

21

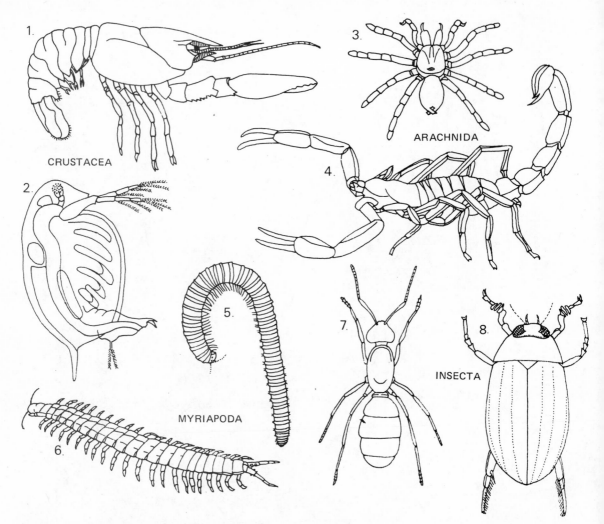

Fig. 3:2 Phylum Arthropoda: representatives of four classes. Crustaceans: 1. Crayfish $\times\frac{2}{3}$. 2. Water flea $\times 25$. Arachnids: 3. Spider $\times 1$. 4. Scorpion $\times 1$. Myriapods: 5. Millipede $\times 2$. 6. Centipede $\times 2$. Insects: 7. Ant $\times 8$. 8. Beetle.

6. ARTHROPODA

The body is segmented and has a hard outer covering (exoskeleton). Jointed limbs are present. This phylum is the largest of all and is divided into four large classes:

a) *Crustaceans*. Arthropods with limbs attached to most segments. Usually aquatic, breathing by means of gills, e.g. crabs, prawns, lobsters, water fleas and wood lice.

b) *Insects*. Arthropods with the body divided into three parts. They have three pairs of legs and usually two pairs of wings, e.g. butterflies, bees, beetles, flies and locusts.

c) *Arachnids*. Arthropods with the body divided into two parts. They have four pairs of legs, e.g. spiders, scorpions, mites and ticks.

d) *Myriapods*. Arthropods with the body divided into a lot of similar segments with one or two pairs of legs to most segments, e.g. centipedes and millipedes.

7. MOLLUSCA

Usually have a shell which is single and often coiled, e.g. snails; double, e.g. mussels; or internal, e.g. octopuses and squids. Their bodies are soft and they have an organ called a 'foot'.

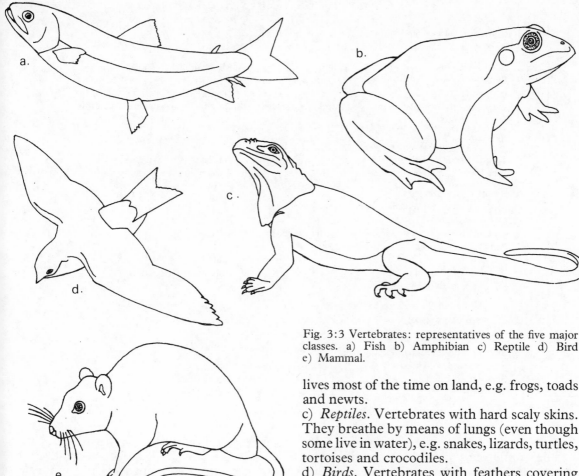

Fig. 3:3 Vertebrates: representatives of the five major classes. a) Fish b) Amphibian c) Reptile d) Bird e) Mammal.

lives most of the time on land, e.g. frogs, toads and newts.

c) *Reptiles*. Vertebrates with hard scaly skins. They breathe by means of lungs (even though some live in water), e.g. snakes, lizards, turtles, tortoises and crocodiles.

d) *Birds*. Vertebrates with feathers covering the body and having two wings and two legs, e.g. ducks, eagles, penguins and ostriches.

e) *Mammals*. Vertebrates with hair on their skins and which suckle their young from mammary glands, e.g. cows, lions, elephants, whales, monkeys and man.

By referring to this classification and the relevant illustrations you should soon be able to put the majority of animals you find into their right phylum, and in the case of arthropods and vertebrates into the correct class. However, a more accurate method is to use a key. Keys are constructed in the form of a series of alternative questions. With your specimen in front of you, you decide which of the first two descriptions fits your specimen best. The one you choose then leads you to the next set of questions, and so on until you end up at the correct group.

8. ECHINODERMATA

Marine animals built on a 5-radial plan as in starfish, brittle stars, feather stars, sea cucumbers and sea urchins. They have suckers called tube feet and they circulate water round their bodies in tubes.

9. VERTEBRATA

This single phylum contains all the animals which have backbones. It is divided into five large classes:

a) *Fish*. Aquatic vertebrates breathing by means of gills, having scales on their skins and possessing fins (no legs), e.g. sticklebacks, trout, eels, and sharks.

b) *Amphibians*. Vertebrates with four legs and scale-less skins, which are usually moist. They usually have a larval stage (tadpole) which is aquatic, and an adult stage which has lungs and

KEY TO CONSPICUOUS TERRESTRIAL ANIMALS

1.	Body not divided into segments and having no limbs	2
	Body clearly segmented *or* having obvious limbs	3
2.	Soft-bodied and slimy, with a muscular foot	Molluscs (slugs and snails)
	Small, shiny and worm-like, with the body pointed at both ends	Nematodes
	Large and snake-like, with the skin covered with scales	Reptiles (snakes and slow worms)
3.	Jointed legs absent	4
	Jointed legs present	5
4.	Worm-like with more than 15 segments	Annelids (earthworms)
	Less than 15 segments	Certain insect larvae
5.	2 pairs of walking legs	6
	3 pairs of true walking legs and usually one or two pairs of wings	Insects
	4 pairs of walking legs+1 pair of leg-like head appendages (pedipalps). Body divided into two main portions	Arachnids
	6 or 7 pairs of walking legs	Crustaceans (woodlice)
	More than 9 pairs of walking legs	Myriapods (centipedes and millipedes)
6.	2 pairs of wings in addition to 2 pairs of walking legs	Insects (a few butterflies)
	2 pairs of walking legs; *or* 1 pair of walking legs+1 pair of wings as in birds and bats	7
7.	Skin with no scales, feathers or hairs	Amphibians
	Skin covered with hard scales	Reptiles
	Skin with feathers	Birds
	Skin with hair	Mammals

The simple key shown above should enable you to put any *conspicuous terrestrial animal* found in Britain into its correct phylum and, in the case of vertebrates and arthropods, into the correct class.

There is no comprehensive key for all the animals and plants you may find. All keys have their limitations because of the number of species involved. However, many useful keys have been constructed which will help you to name many of the animals which belong to the better-known groups. (See Appendix.)

With the help of this key and the illustrations of the main animal groups classify as many terrestrial animals as you can. Here are some ideas on how to find them; make sure that each is placed in a suitable container.
1. Use a beating tray. This is any object like an inverted umbrella or a sheet, held or placed on the ground under a bush or tree; the branches are then tapped sharply with a stick and some of the animals are dislodged and fall into it. The best time to do this is in May, but other months are quite good. Try different kinds of trees and bushes; oak is one of the best. Be careful not to damage the trees.
2. Use a sweep net. This is a tough kind of net which can be swept to and fro amongst grasses and low vegetation. If the contents are turned out on to a black cloth the animals show up better.
3. Scrape together some of the leaf litter which covers the ground in a wood or under a hedge, put it in a large polythene bag and sieve it through a wide-meshed sieve on to a sheet; the animals fall through and can easily be seen.
4. Look in the garden in damp sheltered places such as under large stones or pieces of wood, under compost heaps and under bushy plants and ivy. Many animals seek shelter and moisture during the day. To avoid harming the

Fig. 3:4 Using a beating tray.

Fig. 3:5 Sieving leaf litter.

animals always replace stones and logs exactly as you found them.

You can also collect specimens from streams, ponds, ditches or the sea-shore, according to where you are living, but always remember that small organisms in the daytime usually hide away and so get protection from their enemies; therefore they have to be looked for in the places which give them shelter.

Now let us consider the classification of the plant kingdom. Again, only a simplified system will be used to give some idea of the main phyla.

The main plant phyla

1. BACTERIA
Simple microscopic plants of various shapes with no well-defined nuclei. The majority have no chlorophyll.

2. ALGAE
Plants without roots, stems or leaves. All contain chlorophyll and the majority are green, but sometimes the green is masked by other colours as in the brown and red seaweeds. *Spirogyra* and microscopic forms such as desmids and diatoms also belong to this phylum.

3. FUNGI
Like algae, they have no roots, stems or leaves, but they lack chlorophyll. Some are saprophytes and cause decay, others are parasites living on other animals and plants. They include moulds, mildews, yeasts, mushrooms and toadstools.

4. BRYOPHYTA
Small green plants which usually have leaves and stems but no proper roots. They are placed in two classes:
a) *Liverworts*. Most are small, branched green plants which lie flat on the ground in damp places, often forming large mats.
b) *Mosses*. These usually grow together in cushions or compact masses. They have distinct stems and the leaves have a mid-rib.

5. PTERIDOPHYTA
This phylum includes the ferns, horsetails and club mosses. They have distinct roots, stems and leaves. The ferns produce spores, usually from the underside of the fronds; the horsetails and club mosses produce them from special cone-like structures at the ends of the stems.

6. SPERMATOPHYTA
A very large group containing the majority of familiar plants. They all produce pollen and form seeds.

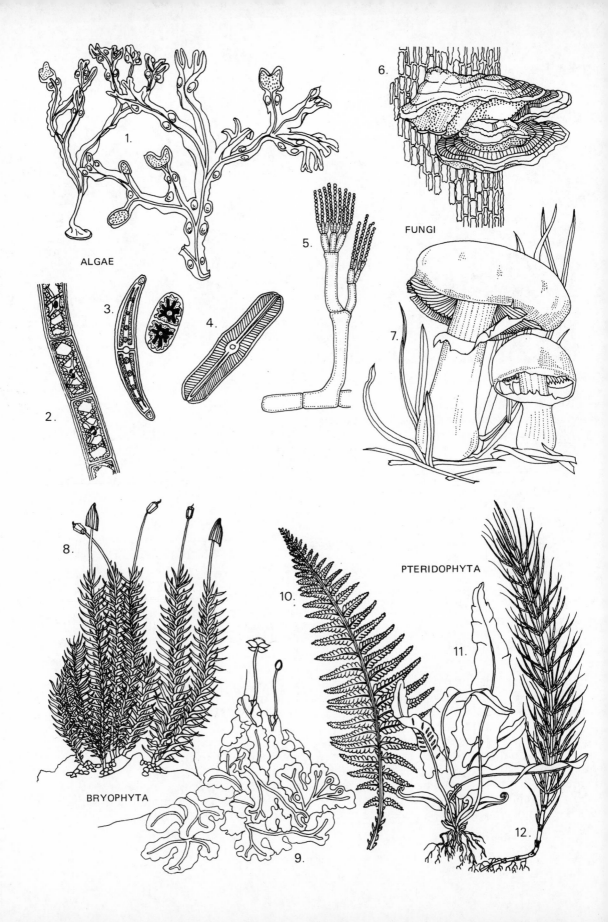

ALGAE

FUNGI

BRYOPHYTA

PTERIDOPHYTA

1.
2.
3.
4.
5.
6.
7.
8.
9.
10.
11.
12.

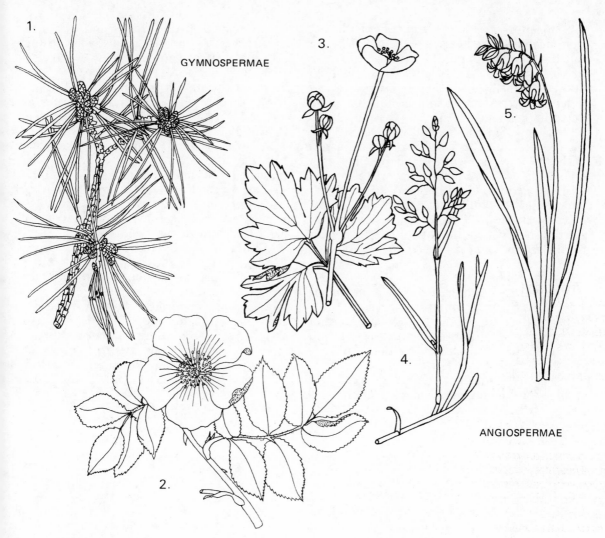

GYMNOSPERMAE

ANGIOSPERMAE

Fig. 3:7 Higher plants (Spermatophyta). Gymnospermae: 1. Pine. Angiospermae: 2. Rose. 3. Buttercup. 4. Grass. 5. Bluebell.

Fig. 3:6 (left) Simpler plants. a) Algae: 1. Brown seaweed $\times \frac{1}{2}$. 2. Filamentous alga $\times 100$. 3. Desmids $\times 200$. 4. Diatom $\times 200$.
b) Fungi: 5. A mould $\times 1000$. 6. Bracket fungus $\times \frac{1}{6}$. 7. Toadstool $\times 1$.
c) Bryophyta: 8. Moss $\times 1$. 9. Liverwort $\times 1$.
d) Pteridophyta: 10, 11. Ferns $\times \frac{1}{8}$. 12. Horsetail $\times \frac{1}{8}$.

27

a) *Gymnospermae*. These bear cones and most have needle-shaped leaves. The majority are called conifers, e.g. pine, larch, fir.

b) *Angiospermae*. All have flowers and produce seeds inside a fruit.

There are two important groups of these flowering plants:

1. Monocotyledons. Usually narrow-leaved plants which include families to which grasses, rushes, lilies, irises and orchids belong. The only common trees in this group are the palms.

2. Dicotyledons. Usually broad-leaved plants which include families to which buttercups, roses, peas, dead-nettles and daisies belong. All the trees which are not coniferous belong to this group, e.g. oak, beech, ash, maple.

You will find it quite easy to recognise members of the Spermatophyta, but get some practice with the others and see how many kinds of ferns, mosses, liverworts, algae and fungi you can find. Here are some suggestions for finding them.

1. You will find algae mainly in water (although one common species called *Pleuro-coccus* occurs on tree trunks and turns them green). Look in ponds, canals, water troughs and rivers. If you are near the sea, remember that all the seaweeds are algae.

2. Liverworts are found in damp places—on the banks of ditches, near drains, at water level where a river goes under a bridge, near waterfalls and on damp ground in woods.

3. Mosses are often found in similar places, but they also occur in drier habitats such as on walls, roofs and tree trunks.

4. Ferns also like the damp; bracken is an exception and does well in dry places. Look in woods, damp hedgerows and shady places; also on old walls.

5. Fungi. You will find the larger kinds in greatest variety in the autumn, but some may be found during any month. Different varieties occur near particular trees, so look in oak woods, pine woods, beech woods, etc. Also look in open fields and on old logs. Smaller fungi are all too common when things go mouldy, and some, like the rust fungi, form bright patches of yellow or orange on the leaves of plants, e.g. groundsel plants in the garden.

4

Some of the simpler organisms

Now that we have a general idea of the classification of organisms we can study some of the groups in more detail.

You will have already noticed that the different phyla of plants and animals were placed in an order which showed a general progression from organisms which were very simple in structure to those which were more and more complex. Thus the algae are relatively simple plants and the angiosperms are complex; similarly, the protozoa are simple and the vertebrates very complex. This orderly progression reflects today the process of change that has been going on in organisms for more than 2000 million years—a process called evolution. (Fig. 4:1.)

The first organisms to exist on the earth are believed to have been composed of the simplest kind of protoplasm, but over immense periods of time they gradually became more complex, giving rise to all the species which have existed in the past and have become extinct, and all which still exist today. However, the rate of change in different organisms has varied enormously; some, having reached a very efficient stage for living in a particular kind of habitat, have hardly changed for many millions of years, while others have gone on changing much more rapidly. As a result, we still have living today plants and animals from different phyla which have retained their relatively simple structure as well as those which are very complex.

Fig. 4:1 Simplified diagram to show how some phyla may have evolved.

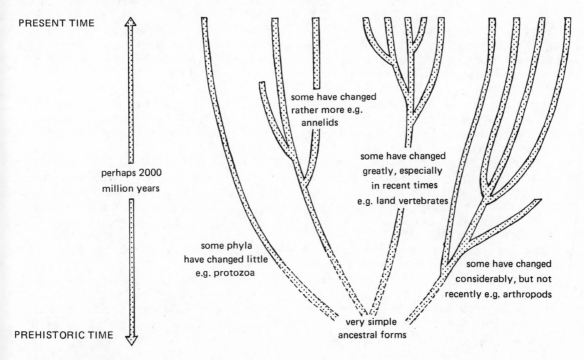

PRESENT TIME

perhaps 2000 million years

some have changed rather more e.g. annelids

some have changed greatly, especially in recent times e.g. land vertebrates

some phyla have changed little e.g. protozoa

some have changed considerably, but not recently e.g. arthropods

PREHISTORIC TIME

very simple ancestral forms

29

In the course of the next chapters we shall be studying a number of organisms from different phyla, starting with the simpler forms and leading to the more complex.

We have already studied two simple organisms—*Amoeba* and *Spirogyra*. Let us now see if we can find some other species which belong to the same phyla. Most are very small so you must look at them under a microscope. You can find them by sampling water from different habitats. You may be given material that has already been collected for you, but it is worth looking in likely places yourself. Here are some suggestions:

1. Look out for ponds, puddles, water troughs or any standing water which looks greenish. By passing a plankton net (Fig. 4:2) through the water many times you can concentrate the contents so that there are more organisms in each drop you examine under the microscope.

2. Take some rotting lawn mowings and place them in a little rain water. Include, if possible, some of the liquid which often oozes out of the rotting material; you should find it teeming with life.

3. If there is a lily pond around, collect a lily leaf which is getting old and beginning to lose its fresh green appearance. Turn it upside down and gently scrape the undersurface with a scalpel, transferring the material to a specimen tube containing a small amount of water. Examine a drop of this.

4. Collect quite a lot of water weed from any pond or slow-moving stream and wash it vigorously, a little at a time, in a dish of water collected from the same habitat. Discard most of the weed, leaving just a few pieces in the water. Allow the sediment to settle overnight and then take samples from the mud at the bottom and examine. Also look at the sides of the dish to see if any *Hydra* are sticking to it. You will be examining these later (p. 33).

You will discover that samples from different habitats contain different species. Some of the organisms you come across will be larger and more complex, such as crustaceans, worms and rotifers. Look at these, too, but do not neglect the small ones with which we are concerned at the moment.

You will be able to classify most of these

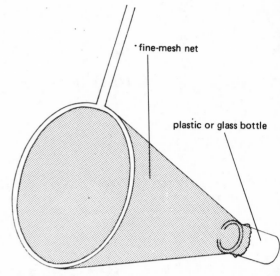

Fig. 4:2 Plankton net.

either into plants (because they are green and have no organs of locomotion) or into animals (because they have no chlorophyll and move actively).

Attempt to classify some of these organisms into their main groups, using the classification below, which lists the main characteristics. Refer also to Fig. 4:3 and other reference books.

1. Flagellates, e.g. *Chlamydomonas*, *Euglena* and *Peranema*. All move by means of one or two threads of cytoplasm called flagella which are very small and difficult to see except under a rather high power. Some are bright green, others are colourless.

2. Ciliates, e.g. *Paramecium*, *Colpidium*, *Vorticella* and *Stentor*. They are surrounded by vast numbers of tiny cytoplasmic hairs called cilia which move them along quite smoothly and relatively fast.

3. Rhizopods, e.g. *Amoeba*. These all move by means of pseudopodia. Some have minute shells. The chalky shells of marine forms, when deposited in astronomical numbers on the bottom of the sea, have formed the chalk rocks of today. By scraping a piece of natural chalk you can see their broken shells under the microscope.

4. Diatoms. These are of many shapes, usually brownish or brownish-green, and have a rigid cell wall, often intricately patterned. Some of them move along very slowly like barges with no visible means of propulsion.

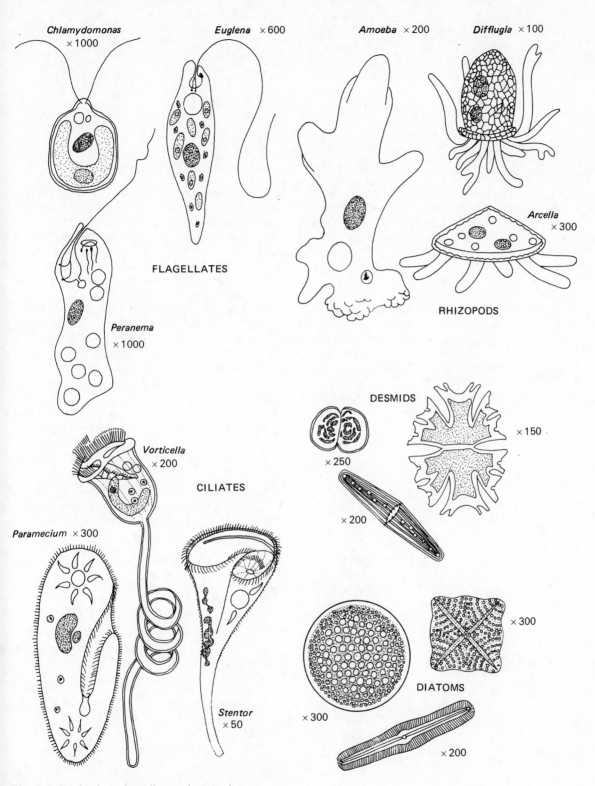

Chlamydomonas ×1000

Euglena ×600

Amoeba ×200

Difflugia ×100

FLAGELLATES

Peranema ×1000

Arcella ×300

RHIZOPODS

DESMIDS

×250

×150

×200

Vorticella ×200

CILIATES

Paramecium ×300

Stentor ×50

×300

×300

DIATOMS

×300

×200

Fig. 4:3 A selection of small aquatic organisms.

31

5. Desmids. These are green and either single or in small colonies. On close inspection each individual appears to be divided internally into two with the nucleus in a transparent region between the two parts.

EUGLENA

These flagellates should be studied in more detail because they combine both plant and animal characteristics. There are many species of *Euglena*. They are found most commonly in ponds or puddles that have been contaminated with organic matter. Sometimes on farms, where manure has got into the water, they are so concentrated that the water becomes bright green. A common species is *Euglena gracilis*, but *E. spirogyra* is larger (Fig. 4:4).

E. gracilis has a spindle-shaped body covered by a non-living, but flexible pellicle, which allows it to change shape to some extent. The cytoplasm contains a number of rod-shaped chloroplasts which give it the green colour; there is a centrally placed nucleus. At the anterior end is a single flagellum which is attached to the base of a spherical reservoir and projects through a narrow 'gullet' into the surrounding water. A contractile vacuole,

Fig. 4:4 *Euglena*: (left) *E. spirogyra*. (right) *E. gracilis* (not to scale).

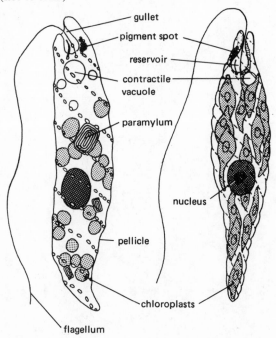

gullet
pigment spot
reservoir
contractile vacuole
paramylum
nucleus
pellicle
chloroplasts
flagellum

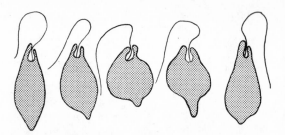

Fig. 4:5 Euglenoid movement. *Euglena* constantly changes shape in this characteristic manner.

which periodically expels its contents into the reservoir, is concerned with osmotic control. Beside the reservoir is a conspicuous orange-red pigment spot which does not detect light itself, but nevertheless plays a part in its response to light. A carbohydrate food reserve called paramylum can be seen as granules scattered within the cytoplasm.

Euglena moves when its flagellum is trailed behind it and thrown into waves; periodically it lashes it from side to side. As it moves forward it rotates on its longitudinal axis. It is also capable of changing shape, passing gradually from an elongated to an almost spherical form (Fig. 4:5).

It reproduces by the simple asexual method of dividing into two longitudinally.

You can find out how *Euglena* responds to light in the following way:

Take some 'green' water containing *Euglena* in a specimen tube and fix black paper round it as in Fig. 4:6, leaving a narrow slit down one side. Place it in moderate light for 24 hours and then carefully remove the paper without shaking the tube. Observe the distribution of *Euglena* in relation to light.

This is an example of irritability, the response of the organism to the stimulus of light. This response is useful because, having chlorophyll, *Euglena* is able to feed rather like a plant by using light and forming carbohydrates from carbon dioxide and water. However, if a colony of *Euglena* is put in the dark (you could try this) they lose their colour and can no longer photosynthesise. Nevertheless, they remain healthy as long as organic matter is present in the water. Without their chlorophyll they are extremely like those flagellates which are *always* colourless and, like them, they feed on

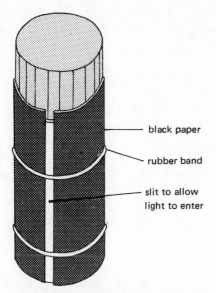

Fig. 4:6 A method of demonstrating the effect of light on the directional movements of *Euglena*.

the organic matter in the water by absorbing it in solution through their surface, thus feeding more like animals. So here is an organism which feeds like a plant in the light and like an animal in the dark. But many biologists think they resemble animals more than plants, because most species are found to need some organic matter to live healthily even when in the light. There are also other characteristics which support this view. Can you think of any?

HYDRA

The animals we have studied so far have all been non-cellular. Now we come to *Hydra*, one of the simplest of the multicellular (many-celled) animals.

The various species of *Hydra*, some green, some brown, belong to the phylum Coelenterata along with the sea anemones, corals and jellyfish. All members of this phylum consist of cells which are arranged in two layers, an outer **ectoderm** and an inner **endoderm**. The body is sac-like having one opening which acts as the mouth, and round it are a number of tentacles.

Hydra may be found in ponds, lakes or streams and may be collected by the method suggested on p. 30. They are easiest to find in summer and autumn, as by then their numbers will have built up after the winter. They attach themselves to water weed, stones or submerged sticks by means of an adhesive disc at the end of the body.

Examine one under the low power of the microscope in a drop of water on a cavity slide (why not a normal slide?). Note how it expands and contracts. What happens to a fully-expanded *Hydra* if you tap the bench near the microscope?

Note the colourless layer of ectoderm cells on the outside of the *Hydra*. The rest of the animal is coloured because the endoderm cells show through. The colour is due to the presence of microscopic algae living within these cells.

Note also the tentacles; how many are there on your specimen? The tentacles surround a small cone at the top of which is the mouth. Only when something passes through it can the mouth be seen.

More detailed structure

The body wall when seen in section (Fig. 4:8) shows that the two layers of cells are separated by a thin layer of jelly, the **mesogloea** (in jellyfish this is a very thick layer, hence the name).

Fig. 4:7 *Hydra* with bud.

The ectoderm and endoderm cells vary greatly in both structure and function, illustrating once more the principle of division of labour between the cells of an organism. The different kinds of cells and their functions are summarised below:

Ectoderm
1. Muscle cells which have contractile tails all lying parallel to the long axis of the body or tentacles. When these contract the body or the tentacles become shorter.
2. Sensory cells which detect touch or other stimuli.
3. Nerve cells which conduct 'messages' from one part of the body to another.
4. Sting cells which are used for paralysing and catching prey.
5. Reserve cells which are capable of growing into other kinds of cells during growth or when sting cells need replacing.
6. Reproductive cells. These are only formed during sexual reproduction.

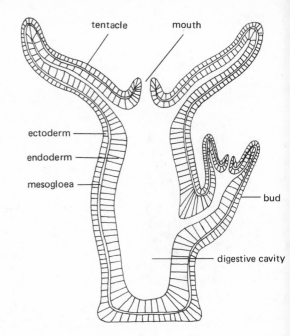

Fig. 4:9 A longitudinal section of *Hydra* with bud.

Fig. 4:8 Types of cells found in the body wall of *Hydra*.

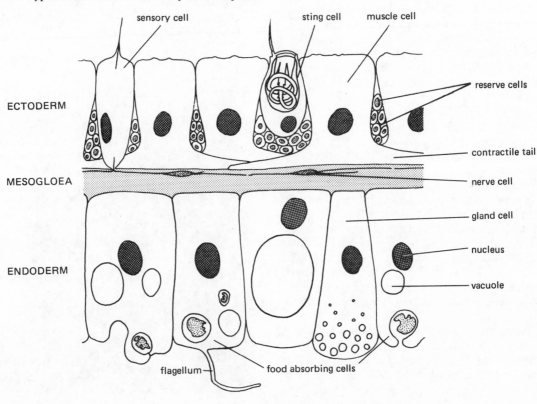

Endoderm

1. Gland cells which secrete digestive juices into the central cavity.

2. Food-absorbing cells. These also act like muscle cells as they have contractile tails arranged in a circular direction which cause elongation of body or tentacles when they contract. When this happens the muscle cells of the ectoderm are relaxed. The food-absorbing cells ingest partially digested food by means of pseudopodia, others have a flagellum which stirs up the digestive fluid in the cavity.

Method of feeding

Hydra feeds on small crustaceans such as water fleas (*Daphnia*).

Watch the method of capture by placing some *Hydra* in a specimen tube. When they have become attached and fully expanded, introduce with a pipette a number of water fleas. In the confined space of the tube they will soon make a capture. Use a lens or micro-projector to see what happens. Note how, to begin with, the tentacles are spread out stiffly like a rigid net, thus covering a wider area and making capture more probable.

When a water flea touches a tentacle, note how quickly it becomes paralysed by the sting cells and how the other tentacles quickly curl round the prey and help to push it towards the mouth.

When the prey is forced through the mouth, which is greatly enlarged during the process, the gland cells of the endoderm secrete their digestive juices and the soft parts are broken up into semi-digested particles. These are later engulfed by the food absorbing cells where digestion is completed in food vacuoles as in *Amoeba*. The skeleton of the water flea which is indigestible is then squeezed out through the mouth and removed.

The sting cells (Fig. 4:10) which paralyse the prey are wonderfully specialised cells arranged in groups on the tentacles and less frequently on other parts of the body. Each has a capsule with a hollow thread coiled up inside a bath of poison. When a water flea touches the trigger which projects from the surface of the cell, the capsule contracts and the thread is shot out, turning inside out in the process. Sting cells

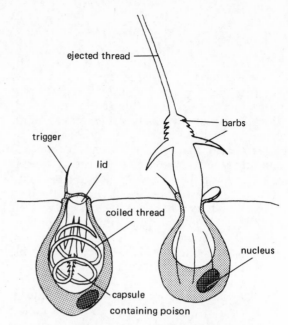

Fig. 4:10 A sting cell before and after discharge.

are of several kinds, but the largest also have barbs able to make a small wound in the prey so that the thread with its poison can enter. Smaller sting cells have no poison, but their threads curl around the bristles of the prey and grip it. All typical coelenterates have these sting cells. If you stroke the tentacles of a sea anemone with your finger it feels rough; this is because it has shot a number of threads into your skin and under a microscope it is possible to see them. Those from sea anemones are not powerful enough to sting you, but those of some jelly fish are. In the Portuguese Man o' War, another large coelenterate, the stings can be extremely painful, and people who bathe in tropical seas need to be careful when they are about.

Fig. 4:11 Locomotion in *Hydra* is usually effected by looping (top); occasionally it carries out complete somersaults (bottom).

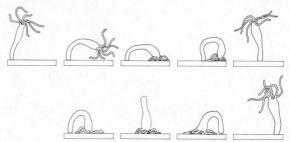

35

Reproduction

Hydra multiplies quickly if kept in a well-lit aquarium and given plenty of water fleas; you can observe the process. First a bump appears on the side and this grows into a bud, which eventually forms tentacles and a mouth at the far end. The cavity of the bud is continuous with that of the parent, so the developing *Hydra* can obtain food from its parent as well as catch its own when its tentacles and mouth have developed. Later the new *Hydra* constricts at the base and pulls itself away from its parent by gripping some water weed with its tentacles. This method of reproduction is asexual and is called **budding**. It occurs constantly throughout the spring and summer when food is plentiful.

Under certain circumstances, especially in the autumn, a sexual method of reproduction takes place (Fig. 4:12). When this happens a group of bumps develop from ectoderm cells in the region below the tentacles; these are the male organs or **testes**. Within them vast numbers of sperms are formed which are shed into the water when the testis wall bursts; they are able to swim with their long cytoplasmic tails. Later on, the same *Hydra* produces one or more swellings further down the body below the testes, and these are the female organs or **ovaries**. Inside the ovary a single egg cell develops, the **ovum**, which gets larger and larger as it stores more food. Eventually the wall of the ovary splits sufficiently to admit sperms, one of which fertilizes the ovum. Now that the egg has been fertilized it can develop into a new individual. By cell division it quickly forms a spherical embryo which secretes round itself a hard horny case which can easily be seen with the naked eye as a dark spot. Later, this drops off into the mud where it remains dormant through the winter months. In the spring the horny case breaks and a new *Hydra* emerges. Thus by forming eggs with protective coats *Hydra* overcomes the problem of winter when the adults die.

Hydra is said to be **hermaphrodite** because a single individual forms both male and female organs. However, self-fertilization is usually avoided as the testes become mature before the ovum is ripe.

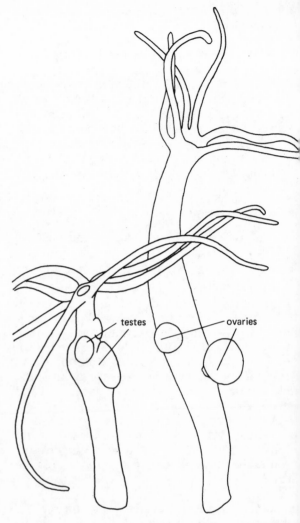

Fig. 4:12 Sexual reproduction of *Hydra*. Although hermaphrodite, the testes usually ripen before the ovaries.

TAPEWORMS

These are parasitic flatworms which live in the intestines of vertebrates, especially mammals and birds. A **parasite** can be defined as an organism which obtains its food from another living organism called the **host** by living on or in it; it is completely dependent upon the host which gets no benefit from the parasite in return. Usually a parasite harms the host to some extent.

A tapeworm's body is long and ribbon-like and divided into very many sections called **proglottides** (sing. proglottis) (Fig. 4:13). It

36

has no mouth or gut, but absorbs food digested by its host through the surface of its body.

The head end is very small and bears suckers, and in some species there are also hooks by which it attaches itself to the wall of the intestine. This is essential as otherwise the muscular movements of the wall of the intestine would squeeze it along with the food and it would pass out of the anus and die. New proglottides are constantly being formed just behind the head while old ones break away from the other end when mature. As each proglottis enlarges it develops a complete set of reproductive organs and by the time it drops off it is no more than a bag of fertilized eggs. These proglottides are deposited with the faeces of the host.

All tapeworms live in two hosts during the course of their life history. The adult stage occurs in the intestine of the **primary host**, and another stage, the **bladderworm**, lives in the muscles of the **secondary host**. We will take the dog tapeworm, *Taenia serrata*, as an example, as this is a common one. The adult stage occurs in the intestine of dogs and foxes, and the bladderworm stage in the muscles of rabbits. When the proglottides, full of eggs, pass out of the dog with the faeces, the proglottides soon die and disintegrate. If the eggs come in contact with vegetation and are eaten by a rabbit, their hard coats are digested by the stomach juices of the rabbit and a microscopic embryo is released from each. This has six hooks enabling it to bore through the intestine wall into a blood vessel. The embryo is carried to the muscles where it grows into a bladderworm. This is a small, glistening, hollow sphere in which the head-end of a new tapeworm is developing.

The bladderworm will not develop further unless the rabbit is eaten by a dog or fox. In this event the head becomes everted and it becomes attached to the wall of the intestine and grows into another worm.

Whatever the species of tapeworm there is always a feeding relationship between primary and secondary hosts, otherwise the bladder-

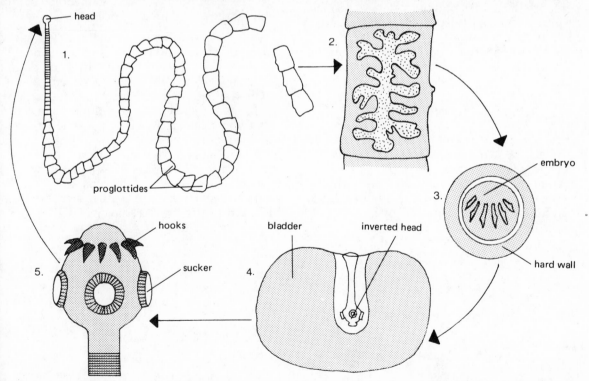

Fig. 4:13 The life history of a tapeworm. 1. Adult worm in intestine of primary host. 2. Single detached proglottis full of eggs which passes out with the faeces. 3. 6-hooked embryo which may be swallowed by secondary host. 4. Bladderworm which develops in the tissues of the secondary host. 5. Head enlarged.

worm would not be able to infect the primary host. Here are some other examples:

PRIMARY HOST	SECONDARY HOST
cat	mouse
lion	buffalo
trout	minnow
rat	rat flea

In spite of this relationship, you will realise that getting from one host to another is a very chancy affair. For example, the likelihood of the dog tapeworm eggs being swallowed by a rabbit is very slight. However, as tapeworms lay enormous numbers of eggs, this makes the event more probable.

It is possible for humans to acquire tapeworms through eating meat which contains bladderworms. One species may occur in pork and another in beef, but the likelihood of infection is extremely remote in countries where proper precautions are taken. If sanitation is good (so that faeces are not left exposed), the chance that eggs will pass from an infected person to the food of pigs or cattle is very small; secondly, if meat is inspected (many countries have strict laws about this), infected meat is unlikely to reach the shops; thirdly, if the meat is properly cooked the bladderworms are killed.

EARTHWORMS

These belong to the phylum Annelida along with marine worms, such as lugworms and ragworms, and leeches.

Most soils contain earthworms although they are most numerous in those which have a high humus content derived from such things as rotting leaves.

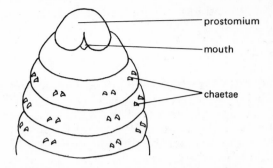

Fig. 4:15 Earthworm: ventral view of head region.

Counting earthworms in a field

How many earthworms do you think there would be in a grassy field such as a playing field? If the area was a large one, e.g. one hectare (10,000 m²), it would obviously be impossible to count them, but it would be reasonable to count the number in a single square metre and multiply by 10,000. But not every square will have the same number. So to be more accurate you should take at least five squares at random and find the average before making the calculation. Proceed as follows:

Use a 2% solution of formalin to bring the earthworms to the surface for counting. As this may temporarily damage the grass do not use your best lawn for the experiment.

Cut a piece of string just over 4 metres long so that when the ends are tied together it will exactly enclose an area of 1 m² when pegged out with four skewers. If the soil is dry, soak it with water first. Now pour over the area 10 litres of the 2% formalin. When the worms appear, collect them with forceps and give each a quick wash in water to remove any formalin, before putting them in a beaker. It will be about 15 minutes before they all come up. Count

Fig. 4:14 Earthworm: external features.

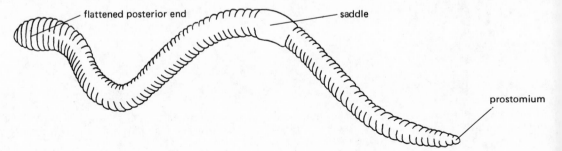

them, take the average for the number of squares treated, and calculate how many there would be in 1 hectare. Keep one large worm for examination, but the rest should be returned to the soil.

It is possible by this method to compare earthworm populations in various kinds of soil and find the conditions they like best. Would it be better to compare numbers or total weight?

Structure

Examine one of the larger earthworms. Notice how the body is divided up into a great number of segments (about 150) most of which are very much alike. If it is mature it will have a **saddle** about one third down from the head; this is used for making an egg cocoon. Examine the head and find the mouth on the underside and the projecting **prostomium** above it (Fig. 4:15). At the opposite end you will see that the body is flatter and wider; this helps the worm to grip its burrow. Right at the end is the anus. Notice the pink line going down the body on the upper surface; this is a blood vessel showing through the skin. Look at this vessel with the help of a lens. Can you see its walls

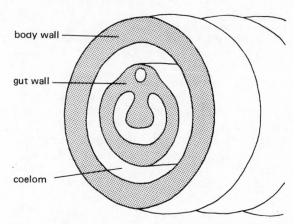

Fig. 4:16 The body of an earthworm is built on a plan of a tube within a tube.

contracting? Which way is the blood flowing? Annelids are the simplest animals to have a blood system.

Gently stroke the underside of the earthworm from the head end backwards; notice how smooth and slippery it is due to the secretion of mucilage from its skin. Now stroke it in the opposite direction. It feels rough because on all but the first and last segments there are four pairs of bristles or **chaetae** which project backwards and help it to grip.

Fig. 4:17 Transverse section through the intestinal region of an earthworm.

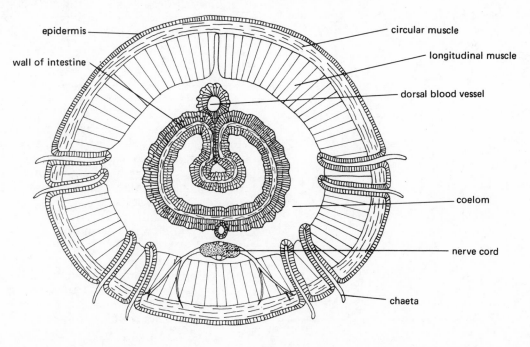

39

Movement

Observe how it moves. You can see this best if you put it on a piece of coarse paper. Notice how it elongates its front end, segment by segment; this would help it to penetrate between loose soil particles when burrowing. Two sets of muscles in the body wall are concerned with movement—one circular, the other longitudinal (Fig. 4:17). When the circular muscles of a segment contract and the longitudinal muscles relax it becomes longer and thinner; when the longitudinal muscles contract and the circular muscles relax it becomes shorter and fatter. Locomotion occurs when waves of muscular contraction pass down the worm (Fig. 4:18). It is prevented from slipping backwards by chaetae, which are protruded from those segments in which the longitudinal muscles are contracted.

You will see from Fig. 4:16 that the body of the earthworm (like all members of the phylum Annelida) is built on the plan of a tube within a tube. The outer one is the body wall which is mainly muscular, and the inner one is the gut. Between the two is the body cavity or **coelom** which is filled with fluid. It acts as a shock absorber and provides something firm for the muscles to act upon.

Fig. 4:18 Diagram to show the sequence of segmental muscle movements during forward locomotion (for details see text).

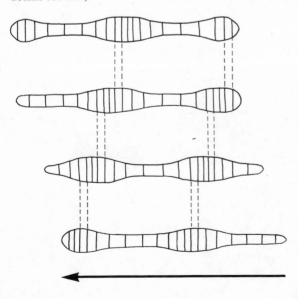

Feeding

Earthworms feed by taking in large quantities of soil through the mouth as they burrow; it is ground up in a muscular part of the gut, the **gizzard**, and digestive juices are poured on it which dissolve the particles of humus. The soluble products of digestion are absorbed into the blood and the remaining soil is discharged through the anus, either in the burrow or as a worm-cast above the surface.

Earthworms increase the amount of humus in the soil by dragging in dead leaves which lie on the surface. You can watch them doing this.

Construct a wormery (Fig. 4:19) and introduce a few large earthworms into it; put some dead leaves on the surface. As earthworms draw in leaves at night, you will have to watch them with a light but this should be very dim.

Also watch the behaviour of earthworms out of doors. Choose a mild, damp night; if it is raining slightly, so much the better. Use a *very* dim torch and look for the worms on a lawn or similar place with short grass. You will see them lying on the surface of the ground with their hind ends in the burrows. Tread very lightly or you will disturb them. Observe how they search for leaves or rotting materials. How do they grip a leaf? Notice the effect of bringing the torch nearer to them. What happens if you tread more heavily? Touch one when it is fully stretched out and see how quickly it moves. This is its reaction if it is attacked by a badger or a little owl. Which muscles is it using when it reacts in this way?

In the daytime earthworms usually remain below ground in their burrows. The latter are often plugged with leaves which they have brought back during the night (on gravel paths they use piles of small stones instead). What purpose does this serve?

There are a number of interesting investigations you could carry out on earthworms. For example, you could test their reactions to different food put out for them; which do they prefer, chopped up onion or cabbage? Do they prefer to drag into their burrows leaves of a particular shape or species? Think out how these experiments could be done.

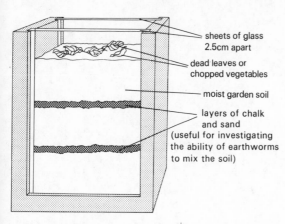

Fig. 4:19 Wormery (the top should be covered with perforated zinc).

sheets of glass 2.5cm apart

dead leaves or chopped vegetables

moist garden soil

layers of chalk and sand (useful for investigating the ability of earthworms to mix the soil)

How do earthworms affect the soil?

Earthworms are of great importance in keeping soil in good condition. By their burrowing activities they make the soil looser and more porous; this improves the drainage and aeration, bringing more oxygen to the roots of plants which need it for respiration. By their feeding activities the soil is broken up into fine particles; it also becomes thoroughly mixed as they take it into their bodies at one level and deposit it on the surface in the form of worm casts, or in spaces within the soil itself. In addition, by drawing leaves into their burrows, organic matter becomes incorporated into the soil, and when this rots, it provides more nutrients for plants.

FUNGI

This important group of plants includes the moulds, mildews, yeasts and toadstools. They differ from typical green plants in two important respects:

1. They have no chlorophyll, and so they are unable to photosynthesise. Instead, they have to feed on complex organic substances (carbon compounds) obtained from dead or living material. If they feed on dead organic matter they are called **saprophytes** and the result of this process is the decay of the material. If they feed on living matter they are called **parasites**.

2. Their bodies are built up from long threads called **hyphae.** Sometimes these form a tangled mass rather like cotton wool called a **mycelium**, which creeps all over and pene-

Fig. 4:20 Moulds growing on damp bread.

trates into the organic matter on which the fungus is feeding. In the larger fungi such as mushrooms, the hyphae become bound together to form the solid reproductive bodies which we use as food. In many of the simpler fungi the hyphae are hollow branching tubes and many nuclei occur in the cytoplasm which lines the wall. In others the hyphae have cross walls and are thus divided into true cells. The walls of the hyphae are not made of the cellulose typical of green plants, but of a nitrogenous compound.

Fungi reproduce by means of spores which are so small and light that they may be carried almost everywhere by the slightest wind current. A mushroom is said to release half a million spores every minute for several days!

Observe these spores by cutting off the stalk of a mushroom and inverting the cap over a piece of paper: put a cover on top to eliminate draughts and examine the paper next day. You will see a pattern corresponding to the underside of the cap composed of millions of spores. If you try this with various toadstools you will find that the spores vary in colour according to species.

Because fungus spores are so common in the air, organic matter which is damp and exposed

Fig. 4:21 Photograph of the inverted head of a mushroom and the spore pattern it has produced.

41

to the air soon goes mouldy as the fungus spores settle, grow and feed upon it. Thus fungi cause decay by feeding on the dead organic matter. Bacteria do the same (p. 189). Decay is a most important process as it causes nutrients to be returned to the soil to be used once more. Imagine what it would be like if the dead bodies of plants and animals never decayed!

You can grow many kinds of fungus for yourself. Take a slice of bread, moisten it with water and then sprinkle some dust on it (not chalk dust). The dust is sure to contain fungus spores. Place it under a glass bell-jar or plastic cover, put it in a warm, dark place, and examine it at intervals over the next few weeks. Note how the mycelia of the different fungi vary in colour and in the way they grow. Look out specially for a species which puts up vertical hyphae which end in black rounded heads. This will probably be either a species of *Mucor* or *Rhizopus*: either will do for examination as their structure is very similar.

Mucor

Examine a portion of the mycelium in a drop of water under the microscope. Note the hollow, branching hyphae. Can you see the cytoplasm lining the walls of the hyphae? Note also the reproductive hyphae whose rounded heads, called **sporangia** (sing. sporangium), give the name Pin-mould to *Mucor*. The sporangia contain the spores which change from pale brown to black as they ripen. Look out for stages in the development of the sporangia (Fig. 4:22). Many will be broken and the spores shed. When this happens naturally the sporangium case splits, leaving only a collar round the hypha. Each of the spores is capable of growing into a new mycelium if it settles on suitable organic matter. This method of reproduction is asexual as no fusion of cells is involved.

Mucor can also reproduce by a sexual method called **conjugation** which results in the formation of **zygospores**: these can withstand longer periods of drought than the asexually produced spores. Zygospores form as a result of contact between two hyphae (Fig. 4:24). The tips swell up, the wall between them

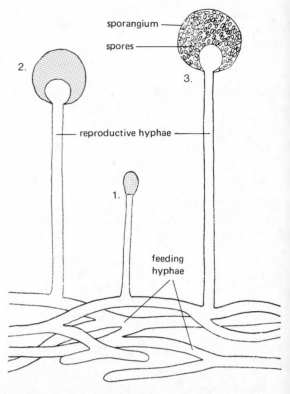

Fig. 4:22 Diagram of part of a mycelium of *Mucor* showing three stages in sporangium formation.

dissolves, and the nuclei fuse in pairs, one from each hypha. The resulting zygospore forms a hard, rough wall round itself which is resistant to water loss and may remain dormant for months before germinating. When this happens, a single vertical hypha grows out from the zygospore and produces a sporangium at its tip. On bursting, many spores are released, each capable of producing a new mycelium.

Fig. 4:23 Photomicrograph of *Mucor* showing zygospores.

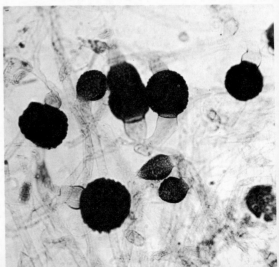

42

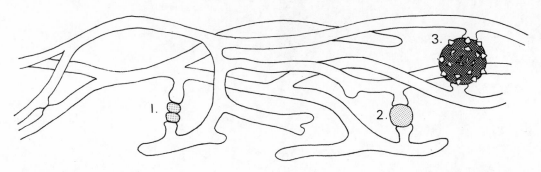

Fig. 4:24 Stages in sexual reproduction of *Mucor*. 1. Hyphal tips meet and swell. 2. Dividing wall dissolves and contents fuse. 3. Zygospore swells and forms hard wall.

Parasitic fungi

In contrast to the saprophytic fungi which feed on dead organic matter and cause decay, there are many parasitic species which live on living plants and animals. One of these, *Pythium de baryanum*, attacks seedlings and causes **damping-off disease**. Its spores are commonly present in garden soil.

Pythium

Examine this species for yourself in the following way:

Put some fresh, damp, garden soil into a pot (Fig. 4:25), scatter cress seeds thickly over the surface and gently press them in. As soon as they show signs of germination, water the soil thoroughly and cover with a polythene bag. Leave in a warm place and note what happens to the seedlings during the next few days.

If any seedlings collapse, remove several carefully, place each on a slide, add a few drops

Fig. 4:25 Method of growing *Pythium*. (Cover with polythene as soon as germination has started).

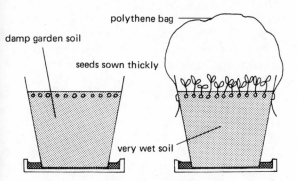

polythene bag

damp garden soil

seeds sown thickly

very wet soil

of lactophenol cotton blue stain and leave for a few minutes. Any fungal hyphae will absorb the stain and turn blue. Now remove the stain with blotting paper, add a few drops of water, put on a coverslip and examine under the microscope. Note particularly the region of the seedling which was at soil level where the stem was weakened, causing it to collapse. Can you see any hyphae stained blue? If so, do they consist of cells? Do they branch? Do they actually enter the cells of the stem? Are there any sporangia present on hyphae which are on the surface of the stem?

Under dry conditions *Pythium* spores remain dormant in the soil, but when it becomes wet and warm enough they become active and swim in the soil water. If they reach a seedling they may enter it through the root hairs and develop into new mycelia which feed on the cells of the seedling causing it to collapse.

Damping-off disease occurs commonly when seedlings of many species are grown under wet, warm and crowded conditions. It is often a menace when raising seedlings in greenhouses. Gardeners prevent infection by using sterilised soil for growing seedlings. The soil is sterilised by heating it strongly; this kills the *Pythium* spores which may be present.

Other fungi which cause diseases

There are vast numbers of other parasitic fungi which cause diseases, and some are of great economic importance. There are, for example, many species of rust fungi which attack wheat, oats, rice and other cereals, causing millions of pounds' worth of damage annually. Scientists are continually carrying

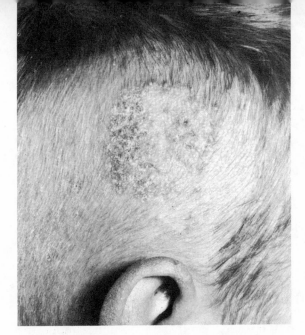

Fig. 4:26 Ringworm: a fungus which attacks human skin.

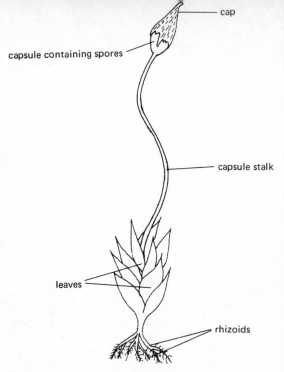

Fig. 4:27 An isolated moss plant.

out breeding experiments in an attempt to produce strains which are immune to these diseases.

One of the mildews, *Phytophthora infestans*, which causes potato and tomato blight, played an important role in history as it caused the complete failure of the potato crop in Ireland in the middle of the last century. This resulted in widespread starvation and caused many families to emigrate to the United States.

The outbreak (1971) of Dutch Elm disease which killed so many trees in southern England was caused by another parasitic fungus. It is passed from tree to tree by a small beetle living in the bark. Another fungus of importance to the forester is the honey fungus, *Armillaria mellea*, a toadstool which spreads underground by means of long black cords made up of hyphae; these enter the roots of trees and may eventually kill them. If a log containing the living mycelium of this fungus is looked at in the dark it will be seen to be luminous.

Some fungi cause diseases in man. Ringworm, in spite of its name, is a disease caused by a fungus which forms a mycelium in the skin causing the hair to fall out, usually in a circular patch. Athlete's foot is a common fungus infection which usually attacks the skin between the toes where it finds warmth and moisture, so it is necessary that feet should be kept clean and the region between the toes

dried after bathing. It spreads very quickly by spores, often on towels; this is a good reason for not sharing them.

MOSSES AND FERNS

Mosses and liverworts belong to the phylum Bryophyta (p. 25). Their structure is intermediate between the algae (which have no leaves and stems) and the ferns (which have leaves, stems and roots). Moss plants do not grow singly, but in clumps or mats.

Examine their structure by separating out individual plants from a clump, washing away any soil that clings to them. If possible choose one which has a capsule attached (Fig. 4:27). Note the presence of simple leaves clustered round the stem and the absence of true roots. The hair-like structures called **rhizoids**, which penetrate the soil, absorb water and help to anchor the plant, but they differ from true roots in having no woody conducting cells.

The capsule which projects from the tip of the moss plant produces the spores. Examine it with a lens. When young it is covered by a cap, but when mature the cap falls off and the spores are dispersed by the wind as they are very light. This only happens in dry weather

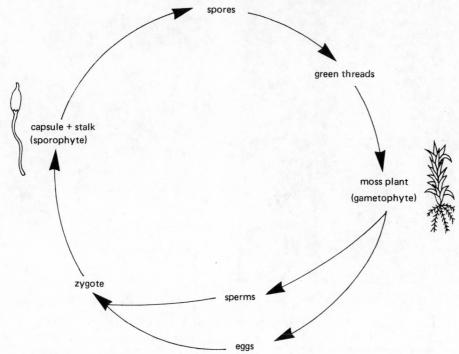

Fig. 4:28 Life cycle of a moss.

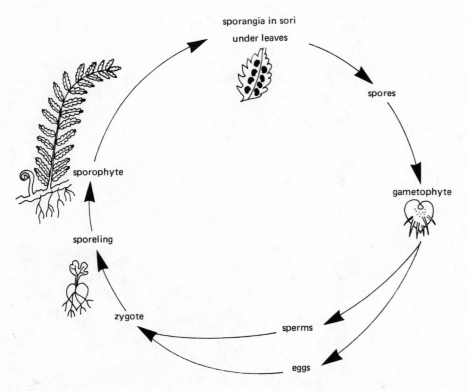

Fig. 4:29 Life cycle of a fern.

Fig. 4.30 Fern plant with fronds uncurling.

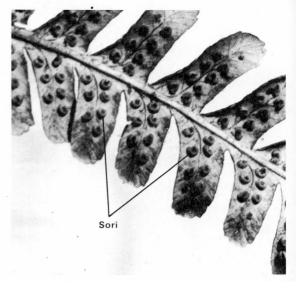

Fig. 4:31 Part of a fern frond showing sori.

because at the entrance to the capsule there is a ring of stiff teeth which close the entrance when wet, but which bend outwards when dry. You can watch this change by altering the conditions. Think out a good technique for doing this.

The life history of a moss (Fig. 4:28) involves two distinct generations which alternate with one another. The first is the typical moss plant with stem, leaf and rhizoids. It is called the **gametophyte** because it forms at its tip the sex organs which produce either eggs or sperms, or in some species both. Fertilization takes place in wet weather, and the fertilized egg (zygote) grows into a capsule on the end of a long stalk. This is the second generation

known as a **sporophyte** because it produces spores asexually. If the spores fall on damp soil when they are liberated, each will form a green thread-like structure from which a new moss plant will develop. This **alternation of generations** between a gametophyte and a sporophyte is characteristic of all mosses and liverworts.

Ferns have a rather similar life history, but in these the sporophyte is the large plant we recognise as a fern, and the gametophyte is a flat green structure less than 0·5 cm long which lives on the surface of the soil (Fig. 4:29). In ferns the spores are formed in sporangia which are grouped together in sori on the underside of the leaflets. The sori appear as brown spots to the naked eye.

Fig. 4:32 Photomicrograph of a transverse section through a sorus showing the sporangia.

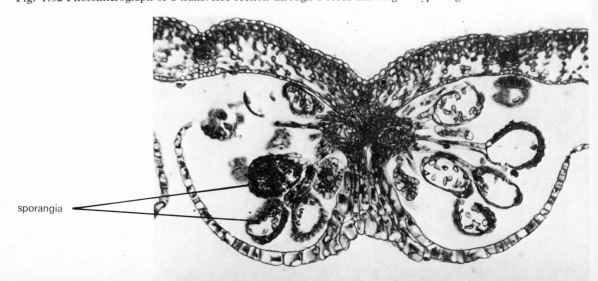

sporangia

5

Organisation of the flowering plant

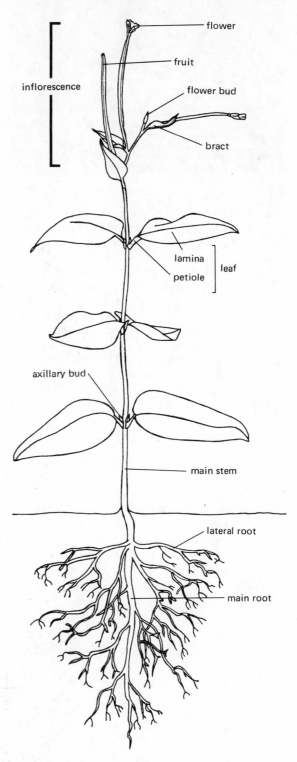

Plants, like animals, have their cells arranged in tissues, organs and organ systems. You can discover the general plan of structure by looking at whole plants of such species as wallflower, shepherd's purse, groundsel or buttercup. They differ in detail, but all have the same basic parts.

Note how each has an underground portion, the **root system**, and an overground part, the **shoot system**. The shoot is formed from stems which bear leaves and buds, and if the plant is mature there will also be flowers and fruits. Compare your specimen with Fig. 5:1 and note similarities and differences.

We can now look more closely at the structure and functions of the different parts.

The root system

Examine some germinating mustard seeds which have been grown on damp filter paper in a covered Petri dish. They show very simple root systems consisting of a main root with vast numbers of **root hairs** looking like cotton wool protruding from the surface. Most root systems start in this way. Later, lateral branches arise with more root hairs near the tip of each, and further branching may occur more haphazardly to produce a very com-

Fig. 5:1 Diagram of a whole plant to show the main structures.

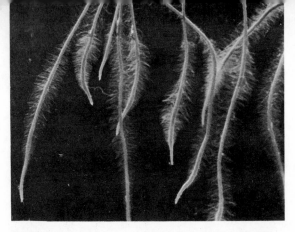

Fig. 5:2 Roots of cress seedlings showing root hairs.

Roots have several functions:
1. They anchor the plant firmly in the ground, preventing it from being washed away by rain or blown over by the wind.
2. They absorb water and mineral salts from the soil and conduct them to the shoot system.
3. They often store food, sometimes getting very large in the process, as in carrots and parsnips.

Roots are admirably adapted for anchorage and water absorption by their pattern of branching. Both fibrous and tap root systems cover a relatively wide area around the plant, making use of the intricate air spaces between the soil particles to grow in all directions. Grasses, because of their fibrous root systems, help to keep the soil particles together, and thus prevent erosion. For this reason they are grown in such places as motorway embankments. A species called marram grass is also planted on sand dunes to help consolidate the sand. It is effective because of its network of underground stems and the adventitious roots which arise from them.

Different species tap different depths of soil, some being shallow rooting, others penetrating deeper. In this way they can live together in the same place without competing too much for the available water and salts.

The root pattern depends to some extent on the depth of the soil. When wheat grows in the very deep soil of the Ukraine its fibrous system may go down as much as 2·7 m, but in Britain this depth is impossible because of the underlying rock. Some trees growing in shallow soil send their roots parallel to the surface in all directions for 30 m or more. What depth of rooting would you expect to find in plants growing in a) deserts, b) sand dunes, c) marshes?

plex system. If the main root remains bigger than the rest it is called a **tap root system** (Fig. 5:3), but when most of the roots appear alike and arise as a bunch from the base of the stem it is called a **fibrous root system**. The latter is characteristic of grasses (Fig. 5:3). Roots can also arise from stems and even leaves; they are then described as **adventitious**. If you pull off a piece of ivy from a tree trunk, you will find it was attached by roots growing from the stem. When gardeners take cuttings, they remove a shoot and roots grow out from the base when it is planted. Similarly, when a leaf of the African violet is planted, roots will grow from the leaf stalk.

Fig. 5:3 1. A fibrous root system. 2. A tap root system.

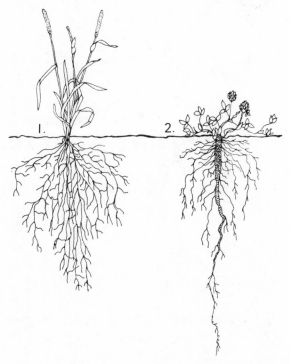

Structure of the root

We can study this better by examining thin sections under the microscope. If you examine a longitudinal section of a root tip (Fig. 5:4), you will see the **root cap**, a protective layer which prevents the delicate cells of the root from becoming damaged as the root grows between the soil particles. Behind it is a region of small cells which look alike and are capable of active division. This is the **region of cell divi-**

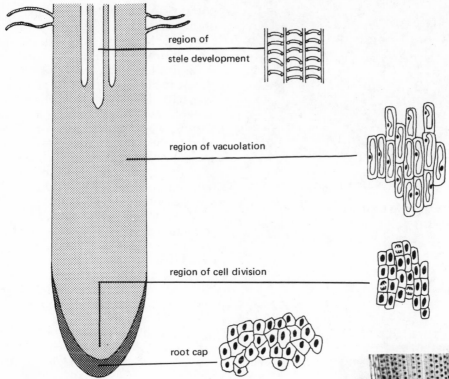

Fig. 5:4 Diagram of a longitudinal section through a root tip showing regions.

region of
stele development

region of vacuolation

region of cell division

root cap

sion. Compare these cells with the ones further back which become progressively more elongated and have large vacuoles. This is the **region of vacuolation**. Still further from the tip you will see that the centre of the root begins to look different from the rest. Here the cells have developed into the main conducting region of the root: this is known as the **stele** or **vascular system**. If you look at a transverse section through a young root, this will give you a better idea of the structure of the central stele (Fig. 5:7). You should see that it is composed of different kinds of cells called **xylem** and **phloem**. The xylem is the woody portion of the root. It largely consists of elongated vessels, which conduct water and salts and are continuous with similar vessels in the stems and leaves. These vessels arise from columns of living cells placed end to end; they become long pipes when the end walls of these cells break down. The side walls are then strengthened with **lignin** (wood) and the cytoplasm eventually dies; hence they can

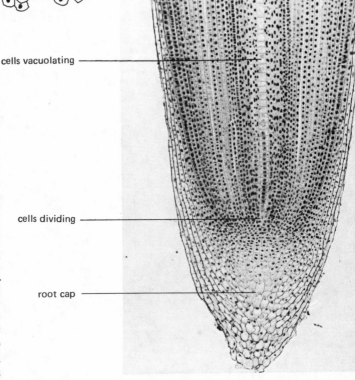

cells vacuolating

cells dividing

root cap

Fig. 5:5 Photomicrograph of a longitudinal section through the root tip of onion (× 45).

49

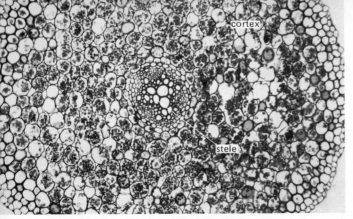

Fig. 5:6 Low power photomicrograph of a transverse section through a young buttercup root.

no longer be called true cells. The phloem, by contrast, consists of living cells. Some of these, the sieve tubes, are also arranged end to end, their end-walls being perforated. This facilitates the conduction of food, made in the shoot, down to the tissues of the root.

The stem

A stem differs from a root in bearing buds and leaves although they are not always easy to see. It ends in a terminal bud which is the growing point of the stem. Other buds arise in the axils of the leaves. When a gardener prunes roses or fruit trees he makes a cut slightly above a bud. This stimulates the bud to grow into a new shoot in the spring.

The majority of stems are rigid and upright, Some, like those of the strawberry, creep along the ground, while others such as honeysuckle or runner beans obtain support by twisting up other plants. Some stems remain underground and are called **rhizomes**. It can be difficult to find the leaves on these because they are reduced to scales which eventually fall off leaving only a scar. Some rhizomes swell up at the ends to form **tubers**, as in the potato; you can tell these are stems by the buds and leaf scars. Various types of stems are illustrated in Fig. 5:8.

The stems of herbaceous plants are green, but in shrubs and trees they become brown due to the formation of a corky bark on the outside.

The main functions of the stem are support and conduction. The leaves have to be supported in such a way that they receive enough

Fig. 5:7 Photomicrograph of the stele region of a buttercup root much enlarged.

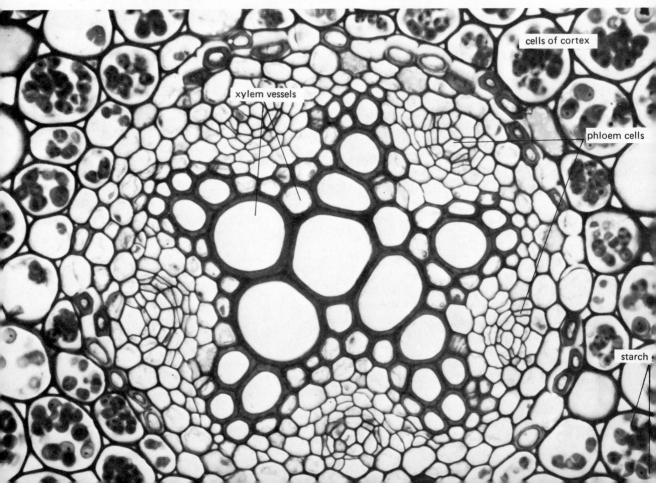

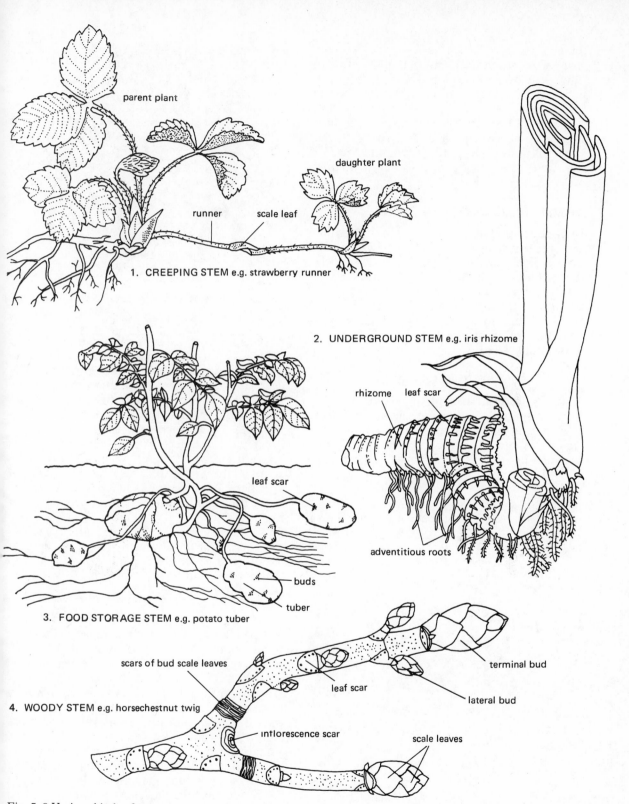

1. CREEPING STEM e.g. strawberry runner

parent plant

daughter plant

runner

scale leaf

2. UNDERGROUND STEM e.g. iris rhizome

rhizome

leaf scar

adventitious roots

3. FOOD STORAGE STEM e.g. potato tuber

leaf scar

buds

tuber

4. WOODY STEM e.g. horsechestnut twig

scars of bud scale leaves

leaf scar

terminal bud

lateral bud

intlorescence scar

scale leaves

Fig. 5:8 Various kinds of stems.

light for making food; this is why the *arrangement* of leaves on a stem is important. The flowers also need support because they need to be in the best position to be detected by insects or, if wind pollinated, to be blown by the wind. The stem also acts as the conducting system between leaves and roots, so that water and salts pass to the leaves and the food the leaves manufacture passes from them to other parts.

Structure of the stem

By examining a transverse section through the stem of a sunflower (Fig. 5:10), you will see that essentially it consists of a tough sheath of cells, the epidermis, a mass of thin-walled cells inside comprising the cortex and the pith, and a ring of vascular bundles. Keeping in mind the main functions of the stem which are support and conduction, let us find out how it is structurally adapted to carry them out.

How does a stem keep rigid?

You will know that if plants are deprived of water they will droop. We can deduce from this that water has something to do with its rigidity. Does the water cause some sort of tension in the stem? These experiments should help you to find out:

Fig. 5:9 High power photomicrograph of a transverse section through the vascular bundle of a sunflower stem.

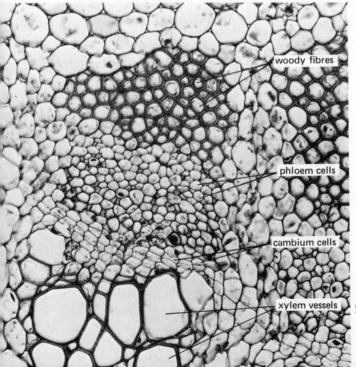

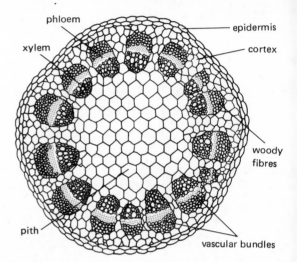

Fig. 5:10 Drawing of a transverse section of a young stem of sunflower.

Take a young green stem; a dandelion stalk will do. Cut it lengthwise into two and cut one strip into two again by a similar cut. Note how each strip curls and see whether the epidermis is on the inside or outside of the bend. What does this tell you about differences in tension between the epidermis and the inner cells?

Now take a piece of rhubarb stalk (this is really a large leaf stalk), and carefully cut a strip of the epidermis, leaving one end still attached. What happens when you try to replace the strip in its original position. Does it fit?

Now take a small cork borer and force it into the cut end of the rhubarb stalk to a depth of about 5 cm. Withdraw it carefully. What do you notice about the cylinder of cells you have isolated in the centre?

From these experiments you should be able to deduce that there are tensions inside these plant organs. You will see later (p. 124) that when cells are given plenty of water they absorb it by osmosis and swell up. This is what the pith cells do, with the result that they press against each other and against the epidermis. Because the epidermal cells are more rigid, they can only stretch a little. Thus the intake of water by the pith, and the pushing outwards of the expanding cells against the epidermis, produce a rigidity rather similar to the inner tube of a bicycle tyre when it is filled with air: it presses against the outer tube which

52

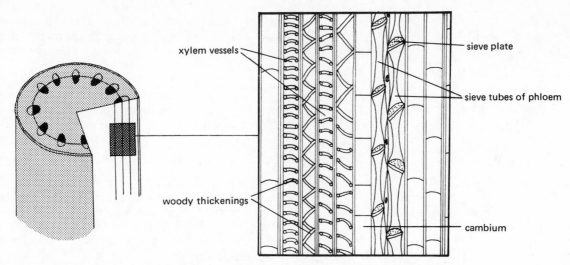

Fig. 5:11 Diagrammatic longitudinal section through a vascular bundle of a sunflower stem.

will not stretch much, so the more air that is pumped in the more rigid the tyre becomes.

Shoots are subjected to high wind pressure which could easily bend and damage them. To withstand this bending strain, they have certain structural adaptations. Look at Fig. 5:10 once more. As wood provides greater support than cellulose, the position of the wood is significant. Can you see that the wood occurs in the form of a ring of girders near the outside? This arrangement is excellent because it prevents too much bending; it is used for the same purpose by civil engineers in constructional work, e.g. the steel used for scaffolding is tubular. Compare this with the root structure where the lignified material is in the centre (as in a cable), an arrangement best suited to withstand a pulling strain.

The vascular bundles not only aid rigidity but contain the conducting elements for water, salts and soluble foods. Each bundle consists of xylem, phloem and **cambium**. The xylem vessels, for water and salt conduction, are continuous with those in the root and leaves; they have smaller, lignified fibres around them to give added support. The phloem, as in the root, consists of sieve tubes for food conduction, and smaller, surrounding cells. Phloem cells have thin cellulose walls which might easily be crushed if it were not for the protection afforded by the xylem on the inside and a tough bunch of woody fibres on the outside.

The cambium, which lies between the xylem and phloem, consists of a thin layer of cells which are still capable of cell division (all the other cells of the stem such as xylem vessels or pith cells are permanent structures). It is by the division of these cells that a stem grows in thickness.

The leaf

Leaves all have a flat green **lamina** or blade, and most of them, a **petiole** or stalk. The petiole is attached to the stem and enters the leaf to form the mid-rib which gives rise to a branching system of veins. However, leaves differ from each other in shape and in many other ways (Fig. 5:12). Some are hairy, others smooth. Some are in one piece (simple), others are divided into complete leaflets (compound). Simple leaves may be partially divided into segments (dissected). The edge of the leaf may be smooth (entire) or toothed (serrated). There may be a petiole present (petiolate) or it may be absent (sessile). Two green projections may occur at the base of the petiole, the stipules (stipulate), or they may be absent (non-stipulate). The veins may arise as branches from the mid-rib forming a minute network which can be seen when you hold it up to the light (reticulate veins), or the main veins may run more or less parallel to each other as in grasses (parallel veins).

Leaves also have characteristic smells and tastes. We flavour food with herbs such as mint or parsley, but some leaves are bitter or poisonous and we leave them alone. Could it

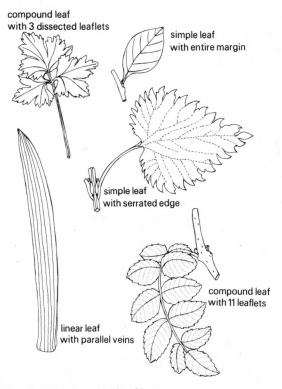

compound leaf
with 3 dissected leaflets

simple leaf
with entire margin

simple leaf
with serrated edge

compound leaf
with 11 leaflets

linear leaf
with parallel veins

Fig. 5:12 Various kinds of leaves.

be that a bad taste or smell would help to protect a plant from being eaten by animals? Can you think of any evidence to support this hypothesis?

Collect leaves from a number of different plants, selecting those which vary widely in their characteristics. Try and describe each as accurately as you can using the descriptions given above.

There are several important processes which occur in leaves. First and foremost there is **photosynthesis**, the manufacture of carbohydrates such as sugar and starch. For this process light is needed, as well as carbon dioxide from the air, and water which comes up from the soil through the xylem vessels of root and stem.

Secondly, **respiration** occurs in all the living cells of the leaf, the oxygen needed for the purpose diffusing in from the atmosphere. Thirdly, **transpiration** takes place. This is the evaporation of water from the leaf, the water

Fig. 5:13 Photomicrograph of a transverse section through part of a privet leaf including the mid-rib.

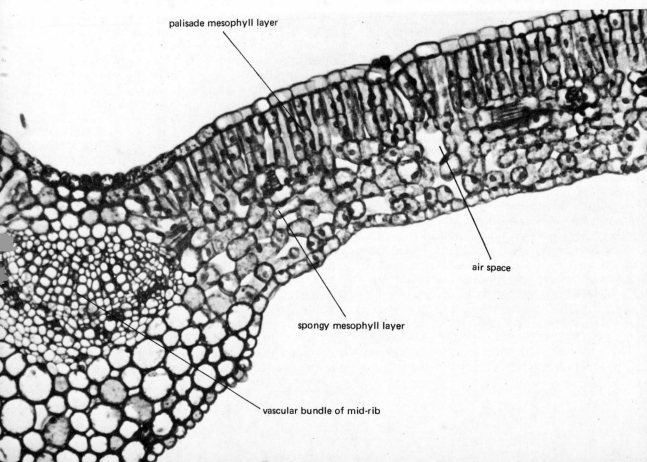

palisade mesophyll layer

air space

spongy mesophyll layer

vascular bundle of mid-rib

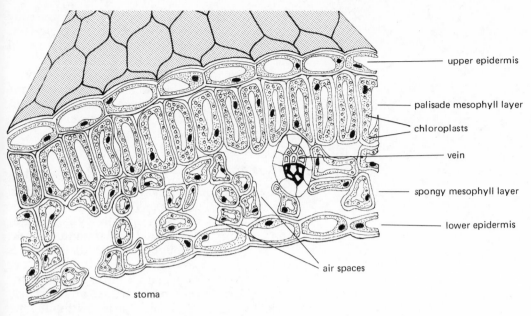

Labels on diagram:
- upper epidermis
- palisade mesophyll layer
- chloroplasts
- vein
- spongy mesophyll layer
- lower epidermis
- air spaces
- stoma

Fig. 5:14 Diagram showing the microscopic structure of a leaf.

vapour passing out into the atmosphere.

In all three processes gases pass in or out of the leaf. Hence the greater the surface area the more efficient the exchange of gases will be. Thus the flat shape of leaves is an adaptation which allows quick gaseous exchange and at the same time allows a maximum amount of light to fall on them for photosynthesis.

Internal structure of a leaf

If you examine a transverse section of a leaf under a microscope (Fig. 5:14), you will see that the epidermis is a layer of colourless cells whose outer walls are thicker than the inner ones. This is due to the presence of a **cuticle** on the outside which helps to reduce evaporation of water and is a protection against mechanical injury, insect attack, or the entry of fungi and bacteria. It also helps to maintain the shape and rigidity of the leaf. In many leaves the cuticle is thicker on the upper surface than on the lower. Can you think of a reason for this?

The lower epidermis is perforated by a vast number of microscopic pores, the **stomata** (sing. stoma); in some species they occur on both surfaces. Stomata allow gases to pass in

and out of the leaf and are capable of opening and closing, thus regulating this gaseous exchange.

Between the two epidermal layers are numerous mesophyll cells which contain chloroplasts. The latter contain the green substance chlorophyll, which is necessary for photosynthesis. The mesophyll cells vary considerably, those near the upper surface are elongated, crowded together and contain many chloroplasts, while those below are more irregular, have larger spaces between them and contain fewer chloroplasts. Why do you think most of the chloroplasts are near the upper surface?

The mid-rib and side veins which can be seen in cross section contain the xylem and phloem elements which are continuous with those of the stem and leaf stalk.

Thus the structure of a leaf is well adapted for its primary function of photosynthesis by having a large leaf surface to receive the light, many chloroplasts in positions where the light will reach them, a system of air spaces for conducting the necessary gases to and from the cells, and veins to bring water to the cells and conduct the sugars made in them to other parts of the plant.

6

Insects: 1

Insects belong to the phylum Arthropoda along with crustaceans, arachnids and myriapods. They share with them the important characteristics of having:
1. A skeleton on the outside, i.e. an exoskeleton.
2. A segmented body.
3. Various paired appendages which are jointed, e.g. legs.

The most successful invertebrates

Biologically speaking, insects are an extremely successful class of animals both in terms of numbers and in the distribution of their species. Nearly a million insects have already been named, i.e. more than 80% of all animal species, and many of these species contain astronomical numbers of individuals. In terms of distribution the picture is equally impressive as they are to be found in almost every kind of habitat except the sea, where only a few highly specialised insects occur.

The insects we shall study have been selected to illustrate aspects of insect life which are particularly important and interesting. Locusts and butterflies have been chosen to illustrate the basic structure of insects and their contrasting life histories. Mosquitoes and flies will be studied largely because of their important role as carriers of disease. Finally, the honey bee has been selected as an example of a species which has evolved a most complex society and which has even developed a 'language' of its own.

The importance of an exoskeleton

The exoskeleton of an insect is called a **cuticle**. It is an excellent protection against mechanical injury. When a blue-bottle fly goes head-on into the glass of a window it is unhurt thanks to the cuticle's toughness. Yet the cuticle is very light and in many ways resembles a very strong plastic. The cuticle also keeps out fungal and bacterial spores which might otherwise damage the living tissues within. However, the chief function of the cuticle is to prevent too much water evaporating from the body, thus allowing insects to live in relatively dry places. The cuticle also aids movement as it serves as a rigid base for the attachment of muscles.

An exoskeleton also brings certain problems; these are rather similar to those which faced mediaeval knights when they wore armour. One is that any continuous hard covering impedes movement. To overcome this difficulty it was necessary for both knights and insects to have joints in parts of their armour. Nevertheless, the joints are the weak spots and in insects may serve as the point of entry for such attacking weapons as stings and poison jaws.

Growth is another problem if the armour does not stretch. The knight had no option but to acquire some new armour; the insect solves the problem by a process called **ecdysis** or moulting. So, when the pressure of growth becomes too great, all insects periodically split their cuticle, having first formed a new soft one underneath to take its place. This new cuticle is able to stretch considerably before it hardens up, consequently, after a moult, an insect looks distinctly bigger than it did a few minutes before. But ecdysis also has its hazards as the animal is very vulnerable before the new cuticle hardens, so many insects hide away during the moulting period.

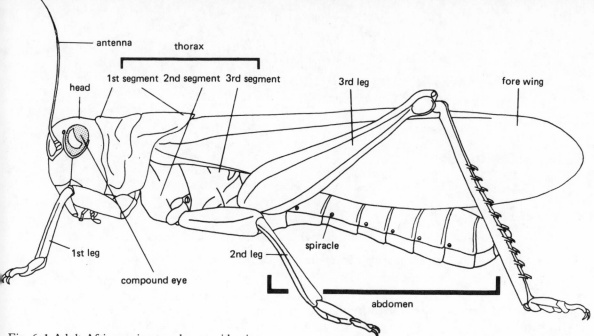

Fig. 6:1 Adult African migratory locust, side view.

LOCUSTS

Locusts are excellent to study as they are large enough for you to see their structure easily (Fig. 6:1), you can breed them without difficulty in the laboratory and you can watch their behaviour at all stages of their life history. In addition, they are of great economic importance.

Locusts, grasshoppers and crickets belong to the order Orthoptera (straight and narrow wings). The African migratory locust, *Locusta migratoria*, lives in hot countries, hence it is best to keep them in heated cages.

Set up a cage as in Fig. 6:2. You should aim at a temperature of about 34°C by day and about 28°C at night. This can usually be done by using two electric bulbs by day and one by night. Put in sticks for the locusts to rest on, and for their main food provide fresh grass each day, with wheat bran to supplement it. Clean the cage daily. When breeding, the adults will need sand in the specimen tubes for egg-laying which should be kept moist by adding a little water each day.

As locusts are very active creatures you will be able to see their structure best by examining a freshly-killed one.

Notice first of all the three arthropod characteristics which, of course, are shared by all insects:

1. The hard exoskeleton. By feeling it you will see that it is both tough and flexible. It not only covers the body, but also forms the main substance of the wings.
2. The segmented body. Note how the posterior segments are all rather similar.
3. The jointed appendages; these are in pairs.

Now study the features which are characteristic of insects. The body is divided into three main parts—**head**, **thorax** and **abdomen**.

Fig. 6:2 Locust cage.

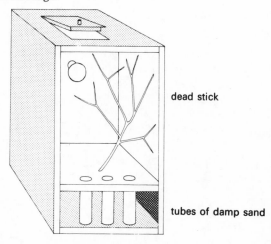

dead stick

tubes of damp sand

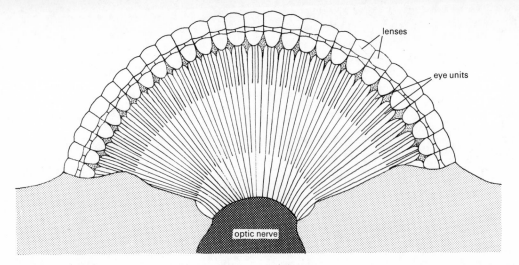

Fig. 6:3 Section through a compound eye (much simplified).

1. *The head*. This bears:
   a) A pair of feelers called **antennae**. Are these jointed?
   b) A pair of **compound eyes**. Look at these under a lens to see the many units of which they are composed. Fig. 6:3 is a section through a compound eye. Each unit points in a slightly different direction, thus the whole eye covers a wide field. The complete image is something like a mosaic of tiny dots of varying intensity, each dot resulting from a single unit.

c) The mouthparts. Insects have complex and very variable structures for preparing and taking in food, but the general plan for all insects is to have:
   (i) an upper lip or **labrum**;
   (ii) a lower lip or **labium**;
   (iii) a pair of jaws, the **mandibles**;
   (iv) a pair of accessory jaws, the **maxillae**.
You can dissect these out with forceps and examine them. Look at the specimen head-on and slightly from below. Look for the two powerful mandibles (Fig. 6:4); they are partly hidden by the labrum, a wide

Fig. 6:4 Head of locust from the front: a) natural position b) with labrum turned back to show the mandi̶̶

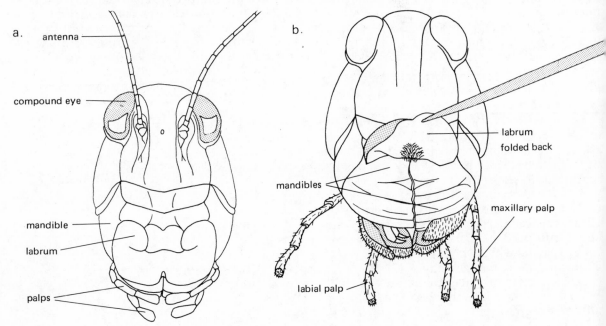

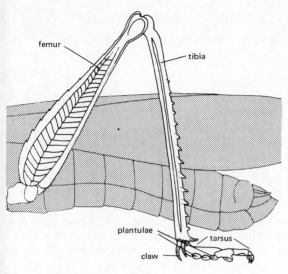

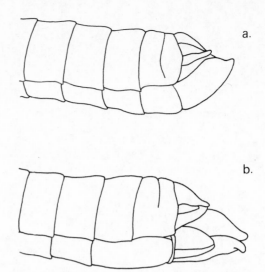

Fig. 6:5 Third leg of a locust.

Fig. 6:6 Abdomen of a) male b) female locust.

plate near the base of the head. Hinge the labrum back to expose the mandibles and carefully remove it with forceps and place it on some wet filter paper for examination later. Now pull off the mandibles. Notice how strong they are and how large the muscles are that have come away with them—in life these help to move them from side to side against each other. The serrated edge of each mandible helps to grind up the food. Now remove the maxillae; each is attached to the side of the head and may be recognised by the long **palp** which forms part of it. Finally, take away the labium which is really a fusion of two structures similar to the maxillae; it has a smaller palp on each side.

Lay out the specimens in their relative positions on the filter paper and examine them under a lens.

2. *The thorax.* In all insects this is composed of three segments; these segments are best seen from the side, as a large plate, the pronotum, covers them dorsally. You will find that the hind pair of wings lie under the front pair. To which segments are the wings attached? Examine the three pairs of legs; one pair is attached to each segment. Notice that each leg is built on the same plan of femur (pl. femora), tibia and tarsus, with two small joints in addition at the body end (Fig. 6:5). Under a lens you should be

able to count the number of joints in the tarsus and see the claws and rounded pads called plantulae. If you move the leg you will notice that the cuticle is softer and more flexible at the joints.

3. *The abdomen.* Of how many segments is it composed? Look at the segment next to the thorax for a pair of rounded membranes, one on each side. These are the locust's ears. They act like our eardrums by vibrating when sound waves fall on them from the air. Locusts make sounds by a process called **stridulation**. They rub the inner side of a back leg (notice the fine spines on it) against the edge of the forewing and the sound which results is amplified because the wing acts as a resonator.

The structures at the end of the abdomen differ in the two sexes (Fig. 6:6); they are concerned with pairing and egg laying. Along the sides of the body you will see a series of pores called **spiracles**. These are the openings of the **tracheal system** which is the breathing system of insects.

The tracheal system

This is a most efficient system of fine tubes called **tracheae** (sing. trachea) (Fig. 6:8). All but the finest tracheae are supported by strengthening rings to prevent them from collapsing; they branch into finer and finer

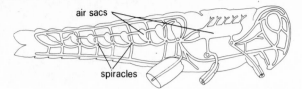

Fig. 6:7 General arrangement of the tracheal system of a locust.

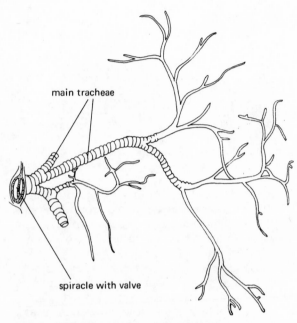

Fig. 6:8 Diagram of a branching trachea.

Characteristics of insects

We can now summarise the main characteristics of insects:
1. The body is divided into a head, thorax and abdomen.
2. There is a single pair of antennae.
3. Normally, two pairs of wings are present in the adult stage.
4. They possess three pairs of legs in the adult stage.
5. All insects breathe by means of tracheae.

Studying living locusts

Now you know something of the structure of a locust you can watch some living specimens and find out the functions of their various organs and how they use them.

1. Observe how they use their legs. Are the back legs used differently from the others?

Fig. 6:9 Male locust dissected to show the main organs.

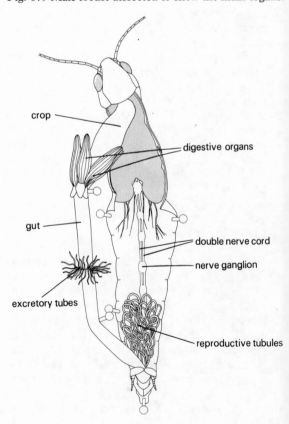

tubes which supply all the organs of the body directly with oxygen from the air. In insects the blood, which is colourless, plays no part in carrying oxygen. The general arrangement of the tracheal system is shown in Fig. 6:7. Each spiracle is controlled by a valve which is normally kept shut, but it is opened periodically to let air in. The movement of air is helped by the alternate expansion and compression of the abdomen which acts as a pump. You see this well when a wasp settles on some jam; its abdomen is constantly pulsating.

Other features can be studied if the locust is dissected (Fig. 6:9). Note especially the nervous system which is ventral in position and consists of a pair of nerve cords joined together in each segment by a ganglion; these are like local brains which control the activity of each segment. The ganglia are themselves to some extent under the control of the brain which is in the head region.

2. Why do you think the femora of the back legs are so fat and long?
3. When they climb up the side, how do they grip?
4. When they are on a twig, what part is played by the tarsus with its plantulae?
5. If they are disturbed, watch how they take off and use their wings.
6. Test their reactions by putting your hand into the cage from the top a) very slowly indeed, b) more rapidly. Remember, in the wild they are attacked by animals which move rapidly.
7. Watch how they feed. Look for the movement of the mouthparts, especially the labrum and jaws. What are the palps doing?

Life history

Apart from differences in the structures at the end of the abdomen, adult males are smaller than the females and are more yellow when mature. Mating takes place when the male climbs on to the back of the female and curves its abdomen round to make contact with the end of the abdomen of the female. They remain together for some time while sperms are being passed from the body of the male into that of the female, a process known as **copulation**. The sperms are stored in a special sac near the end of the egg tube. During egg-laying the eggs move down the egg tube and are fertilized as they pass the opening of this sac.

You may be fortunate enough to see a female laying eggs in one of the specimen tubes containing damp sand. First the locust buries its abdomen into the sand, using the valves at the end to push away the sand and make penetration easier. During this process the abdomen elongates and eventually reaches a depth of about 10 cm. It is then slowly withdrawn and the hole is filled with a frothy material in which up to 200 eggs are laid. The froth quickly hardens to form a protective case round the eggs.

The eggs hatch in about 11 days when they are kept at 34°C (longer at lower temperatures). The young locusts push their way out of the sand and their colour darkens as their cuticle becomes harder; soon they will hop about the cage and start eating the bran or grass. They have no wings and are now called **hoppers**.

They grow quickly and moult their cuticle at intervals. When this is about to happen, they become less active and cling firmly to some object like a twig. The cuticle splits down the back and the locust pulls itself out of the old cuticle. Each stage between moults is called an **instar**. The first instar is the newly hatched hopper, and there are five instars in all before the last ecdysis when the adult emerges. These immature locusts are also known as **nymphs**.

Watch for differences in structures between the various instars, noting especially the ways in which the wings develop from buds in the thorax region. The table below summarises the more important points. Note that the time given is the average time when kept at 34°C with plenty of food.

STAGES IN THE LIFE CYCLE OF THE AFRICAN MIGRATORY LOCUST

Stage		Length	Duration of stage	Main characteristics
Egg		0·5 cm	11 days	Position of eyes visible
1st instar		0·9 cm	5	No wings. Black
2nd instar		1·2 cm	4	Wing buds just visible
3rd instar		1·9 cm	4	Wing buds point down
4th instar		2·3 cm	5	Wing buds point up
5th instar		3·2 cm	8	Wings half the length of body
Adult		5·5 cm	Several weeks	Wings longer than body

61

We can summarise the life history of the locust by saying that there are three main stages—egg, nymph and adult. When the nymph hatches from the egg it already resembles the adult in many ways and its further development is a very gradual one. Therefore, the locust is said to exhibit **incomplete metamorphosis**. This type of life history differs considerably from that of insects such as butterflies, mosquitoes and bees, as we shall see later, and is one important factor used in classifying insects.

The locust as a major pest in the tropics

Locust swarms are a great scourge over large areas of Africa, the Middle East, India and South America. It is difficult to imagine how large these swarms can be. They have been known to cover an area of up to 1,000 square kilometres, and to contain perhaps 40,000 million locusts! Even an average swarm may cover 128–256 square kilometres and contain over 5,000 million locusts. When they fly over, the whole sky is darkened. When they settle they utterly destroy the green vegetation, crops are ruined and famine often follows in their wake. You can get some idea of the destruction they cause when you realise that an average adult of 2–3 g eats its own weight of green food every day. Multiply that weight by 5,000 million!

The African locust, *Locusta migratoria*, occurs in two distinct forms which differ both in appearance and behaviour. The first is the **solitary form** which behaves like a large grasshopper and does relatively little damage, and the second is the swarming or **gregarious form**.

Up to 1921 these two forms were considered to be different species, but it was then discovered that if locusts of the solitary form were bred together under crowded conditions their progeny developed into the gregarious form.

In the wild the solitary form is the usual one, but swarms may arise if a season when conditions are good for the locusts is followed by a bad one. During the good season the solitary locusts breed rapidly, but if the country then becomes scorched up, the food dwindles and the locusts congregate in the few areas where

Fig. 6:10 Part of a locust swarm, Ethiopia.

there is enough water to keep the vegetation green and where reproduction can take place. Thus they become crowded together, breed rapidly and produce a vast swarm. With further reproduction, astronomical numbers may result.

You may wonder how the solitary form gives rise to the gregarious when bred under crowded conditions. The answer lies in a **hormone**, a chemical substance which is produced by a gland in the head of the locust. Under crowded conditions this hormone is secreted in larger quantities and this has the effect of speeding up reproduction and bringing about the changes in the progeny which are characteristic of the gregarious form.

For centuries man has attempted to control locust swarms. Success today lies in spotting the swarms when they are in the hopper stage and unable to fly; they can then be sprayed from the air with insecticide. Hence for control to be effective early information about the location of developing swarms is essential and then quick action must be taken to deal with them. To cope with the problem, several international control organizations have been set up. They receive the information, decide on the strategy and co-ordinate the action. In this way considerable success has been achieved, but locust swarms are no respecters of national boundaries and international co-operation is essential. In some areas this has yet to be achieved.

BUTTERFLIES

These insects belong to the order Lepidoptera (scaly wings), because their four wings are covered with tiny scales of various colours which give them their pattern.

Their life history differs from that of the locust by having a **complete metamorphosis**. This means that there is a rather sudden change between the larval and adult stages. So there are four stages in the life history: ovum (or egg), larva (or caterpillar), pupa (or chrysalis), and imago (or adult).

The term **larva** is used to describe a stage in the life history of *any* animal which lives an independent existence from the adult and differs markedly from it in appearance. Thus the caterpillar does not look like the butterfly and feeds quite differently.

The term **pupa** is given to a resting stage between a larva and an adult. Although inactive, great changes take place internally as larval features are lost and adult characteristics develop. These changes are not very evident externally until the pupal skin splits and the adult, in this case, the butterfly, emerges.

The species we shall study to illustrate this life history is the large white, *Pieris brassicae*. There are two other species which resemble it rather closely (Fig. 6:12): the small white, *Pieris rapae*, and the green-veined white, *Pieris napae*. You will notice that they have all been classified in the same genus because of their similarities. These three species differ mainly in size, the colour of the wing veins underneath, and the appearance and food preferences of the caterpillars. The large and small white both feed on cabbages and cause quite a lot of damage, but the green-veined white feeds only on wild plants of the cabbage family and does no damage to crops.

Large white butterflies are best looked for in gardens, allotments or fields, especially where cabbages and allied species are grown, but they may visit flowers in any situation.

Fig. 6:11 Maize damaged by locusts.

This species has two broods in the year—May and June, and August and September. In some years large numbers migrate from the continent to Britain and swell the ranks of the resident population.

Life history of the large white

a) *Egg laying*. It is possible to make the females lay eggs in captivity. They should first be fed on sugar solution (p. 67), and then placed in a large muslin cage in which they can fly freely. They should be supplied with fresh cabbage leaves and the cage should be placed out of doors in the sunshine. If you watch them in a vegetable garden you will see how they choose members of the cabbage family on which to lay their eggs in preference to other plants. They recognise the plants largely by scent. Sometimes they will lay on nasturtium leaves instead of cabbages and these have been found to contain a chemical substance similar to that found in cabbages. A female usually chooses the underside of one of the younger leaves, laying the eggs, one at a time, to form a compact mass. The number depends

Fig. 6:12 Three closely related species of the genus *Pieris*.

on whether or not it is disturbed while egg-laying, but anything up to 100 can be laid.

The egg is elliptical, but drawn out at its apex, and is 1 mm high. Its flat end is glued to the leaf by a sticky secretion. When first laid it is pale yellow in colour, but soon changes to bright yellow and then to orange.

Examine some eggs under the microscope and notice the pattern on their shells. Each egg contains a supply of food, the **yolk**, on which the developing caterpillar feeds. Perhaps you will see the caterpillars hatch out. They do this when 6–10 days old according to the temperature. You can tell when they are about to hatch as the egg turns a dull colour with a dark region at the top; this is the head of the young caterpillar showing through. Notice how they use their jaws to enlarge the hole at the top of the egg before crawling out. What happens to the rest of the egg shell?

b) *The caterpillar.* On hatching, the tiny caterpillars form threads of silk over the surface of the slippery leaf; this gives them something to grip. The silk is formed from a tubular **spinneret** behind the jaws; this secretes a fluid which hardens into a thread of silk when in contact with air. The caterpillars keep close together at first. They feed by scraping off the epidermis of the leaf with their jaws. Later, as they get bigger, they make large holes in the leaf, working methodically along a cut edge. The caterpillar moults four times during its growth over a period of about a month. Its colour remains the same throughout development. It has three yellow longitudinal stripes, one dorsal and one each side on a greyish green background. It is speckled with black.

Examine first a preserved specimen of a full-grown caterpillar. Note the shiny black head, the three segments just behind it which form

Fig. 6:13 Eggs of the large white butterfly.

Fig. 6:14 Caterpillars of the large white on a cabbage.

the thorax, and the abdominal segments behind these. How many segments form the abdomen? Note the three pairs of true legs attached to the thorax and the fleshy prolegs on the abdomen. Use a hand lens to see their structure. What differences can you make out between them?

You will see that the skin is warty. What is the relationship between the warts and the hairs? Look for the spiracles through which it breathes. There is one pair used for breathing on most segments. Which segments have no spiracles?

Look at the head from the front. Note the powerful jaws and the pair of very small antennae at the side. There are no compound eyes, but there are six pairs of simple eyes which consist of a single lens with a few light-sensitive cells. They are probably just sensitive enough to distinguish light from darkness.

Now observe a living caterpillar and try to find out:

1. The function of the true legs. How does it use them?
2. The function of the prolegs.
3. How the true legs and prolegs are adapted to carry out these functions.
4. How it uses its mandibles. How does the method compare with that of the locust?
5. How does it move? Do you see any similarities to earthworm movement? From your observations, can you deduce how the muscles are arranged in each segment?

c) *The pupa.* The full-grown caterpillar may pupate on the cabbages in the summer, but usually it leaves the food plant and moves rapidly over the ground until it comes to something like a fence, tree or wall up which it can

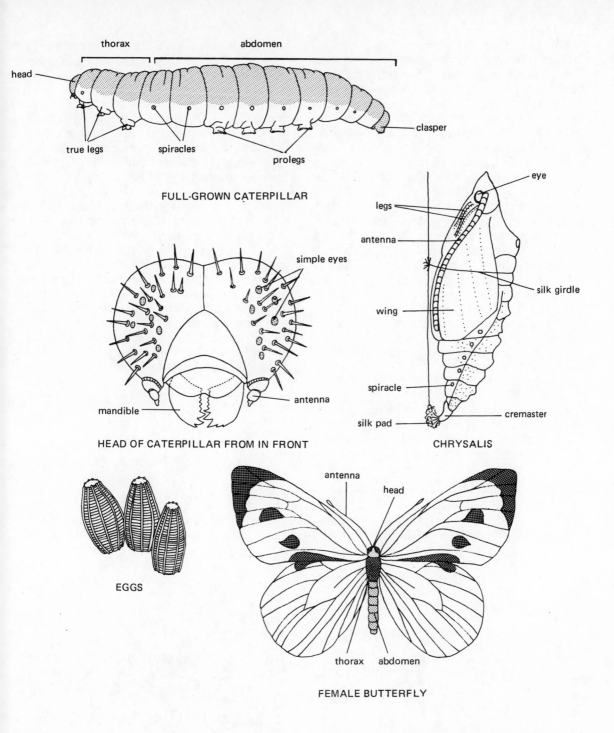

thorax   abdomen

head

clasper

true legs   spiracles

prolegs

FULL-GROWN CATERPILLAR

simple eyes

mandible   antenna

HEAD OF CATERPILLAR FROM IN FRONT

eye

legs

antenna

silk girdle

wing

spiracle

silk pad   cremaster

CHRYSALIS

EGGS

antenna

head

wing

thorax   abdomen

FEMALE BUTTERFLY

Fig. 6:15 Stages of the life history of the large white butterfly.

65

climb. It then selects a place where it will be protected from extreme conditions, and spins a mat of silk into which it digs its claspers. Finally, it forms a girdle of silk round its body which helps to hold it in a fairly vertical position, head upwards. Making the girdle is quite an acrobatic feat, as the silk comes from the spinneret in the head and has to be fixed to the support on either side of the body.

The caterpillar remains quiescent for a day or so, becoming shorter and fatter. Then it starts to wriggle violently and the cuticle splits in the thorax region. Next it frees its head from the old skin and pushes the latter through the girdle by rhythmic movements of the body until the old skin drops off the end. The new pupa now firmly attaches itself to the silken pad with the help of its hooks and is held in its upright position by the silk girdle.

The cuticle of the pupa is at first soft and pale in colour, but it soon hardens and the colour becomes much more like that of its immediate surroundings.

Examine a pupa with a hand lens. Note the butterfly features that are visible (Fig. 6:15). You should be able to make out the wings, legs, antennae and proboscis (feeding tube) below the pupal skin. These structures start to form towards the end of the caterpillar phase.

During the pupal stage further internal changes take place before the butterfly em-

erges. This stage lasts about a fortnight in the summer, but for the second brood, it lasts all the winter.

d) *The butterfly.* The emergence of the butterfly from the pupa is another instance of moulting. The cuticle splits in the thoracic region and the butterfly levers itself out with its legs, and if possible, hangs upside down. The wings are at first small and crumpled, but they are gradually 'pumped up' by blood which is forced into their veins. Once the wings have reached their full size, the blood hardens in the veins and thus gives the wings support. Butterflies bred in captivity sometimes have deformed wings because they have not been given supports from which they can hang when the wings are expanding and hardening.

The female mates soon after emergence, if the weather is sunny. The male is attracted to the female in the first place by sight. After a brief courtship flight they join their abdomens together and sperms are passed into the female's body, as in locusts.

---

1. First examine preserved butterflies of both sexes and note the differences in the wing patterns.

Note how the body is divided into three distinct regions, head, thorax and abdomen, with the legs and wings attached to the thorax.

Examine the head with a hand lens. Note the knobbed ends of the antennae. This is a characteristic feature of butterflies. The antennae are used for touch and smell. Note the

Fig. 6:16 Large white: (left) caterpillar about to pupate (right) chrysalis.

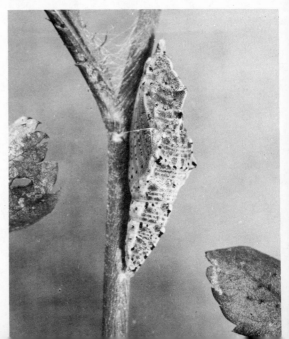

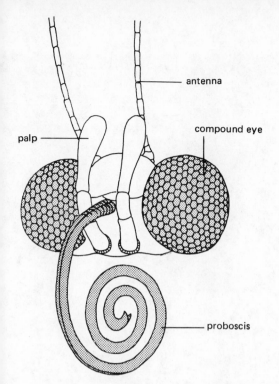

Fig. 6:17 Head of butterfly.

maxillae which are grooved along their inner surfaces and fused together to form a tube, through which nectar is sucked up.

Now examine a wing under the microscope and observe how the overlapping scales make the pattern.

2. Watch how a butterfly feeds. You can see this happen when it settles on a flower. Notice how the proboscis is used.

You can also watch the process in the laboratory by feeding a butterfly yourself. First place a drop of sugar solution on a piece of rough card. Wait until the butterfly folds its wings together above its back. Now hold it carefully by the top of its wings between your finger and thumb. Place the butterfly on the card with its head near the sugar solution, but do not let go yet. It may put out its proboscis immediately, but if not, put a pin into the coil of the proboscis (you will not hurt it), and draw it out until the end reaches the sugar. The butterfly will at once start sucking and you can then release your grip. Gradually the drop will get smaller as it continues to feed.

large size of the compound eyes. They form distinct images of near objects and are quick to detect movement. The mouthparts are very different from those of locusts although they are formed from the same basic parts. The main structure is the proboscis which you can see curled up like a watch spring under the head (Fig. 6:17). It is formed from the two

You can now refer to Fig. 6:18 which summarises the life history.

Controlling numbers

The large white and, to a lesser extent, the small white are the only species of butterfly in Britain which are pests. In large numbers their

Fig. 6:18 Life cycle of the large white butterfly. This species has two broods during the year.

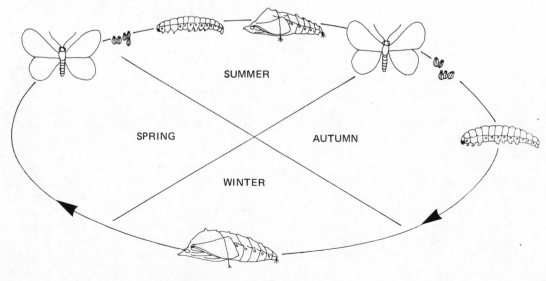

caterpillars can reduce a patch of cabbages to skeletons, leaving only the stalks and leaf veins. When cabbages are grown commercially the damage is prevented by spraying them with insecticides, but it is important that a non-persistent type is used. In a garden the plants should be examined *each week* during the egg-laying season for patches of eggs, which can then be crushed. There are also safe sprays which may be used.

The caterpillars have many natural enemies which help to keep down numbers; birds assist the gardener in this way. A species of ichneumon fly, *Apanteles glomeratus*, is also very effective. The female ichneumon has a sting-like, egg-laying tube, the **ovipositor**. Having found a caterpillar, it uses this ovipositor to pierce the skin, and then passes eggs through the ovipositor into the body of the victim. The eggs hatch into larvae which feed first on the less vital tissues of the caterpillar, which surprisingly, continues to feed and grow. When the caterpillar crawls away to its pupation site most of its organs have been consumed and it soon dies. The ichneumon larvae then eat their way through the skin and form a cluster of sulphur yellow cocoons in which they pupate. Later, they hatch into adult ichneumons ready to attack the next generation of caterpillars.

Another insect called a Chalcid wasp, *Pteranotus puparum*, searches out healthy caterpillars which are about to pupate. It waits nearby until the old cuticle is shed and then punctures the new, soft cuticle with its ovipositor and lays its eggs inside the pupa, with a similar result.

The large white caterpillars are also subject to diseases, so with all these hazards the likelihood of an egg eventually turning into a butterfly is not very great. It has been calculated that for every 1,000 young caterpillars only about 3 become butterflies. This is an example of how the numbers of animals are regulated by natural means.

If you find clusters of yellow *Apanteles* cocoons on the dead skin of a caterpillar stuck to the fence round a cabbage patch, collect several lots carefully and keep them to see if the ichneumon flies come out. You may get a surprise, for it is possible to discover three entirely different species of insect emerging!

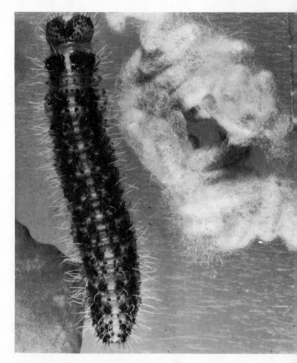

Fig. 6:19 Clusters of *Apanteles* cocoons near the body of the dead caterpillar.

Apart from *Apanteles* there is another ichneumon fly, *Hemiteles glomeratus*, which actually lays its eggs in the bodies of the *Apanteles* larvae by stinging them through the skin of the large white caterpillar. But this is not all, because there is a Chalcid wasp which does the same to the *Hemiteles*! So we have parasites living on parasites and one, two or three species could emerge from those *Apanteles* cocoons!

Butterflies in Britain

There are 65 species to be seen in Britain, although some are extremely rare; others used to occur, but are now extinct. Many species are far less common than they were, and this is probably due to the increasing use of insecticides, the great reduction in marginal land, and the destruction of the specialised habitats in which they live. Some species vary greatly in number from year to year because they migrate from other countries. Some years they fly to Britain in large numbers. The painted lady comes from as far as the Sudan and the clouded yellow from the Mediterranean region.

These two species cannot survive a British winter, but other migrants such as small tortoiseshells and large whites merely swell the numbers of those which are native. The numbers of any one species also fluctuate widely because of seasonal conditions; a wet season, for instance, encourages disease and so fewer survive.

Butterflies should not be collected just for the sake of it, but breeding them is very interesting and worthwhile. This enables the caterpillars to be guarded from their natural enemies and larger numbers of the butterflies can be released. In this way you can help conserve these beautiful insects.

All butterflies have the same stages in their life history, but the details are different. Some have one brood during the year, others up to three. Some pass the winter as eggs, others as larvae or pupae, and a few hibernate in the butterfly stage. Small tortoiseshells, peacocks, commas and brimstones are the more common hibernators; they are usually the first ones to be seen in spring. Butterflies all feed as adults on liquid food; this is usually the nectar from flowers, but sometimes they feed on the juices from rotting material such as fruit. The caterpillars all feed on plant material, some being restricted to a single species, others using several.

Moths are very similar to butterflies, and in practice there is no perfect way of distinguishing them. However, it is true to say for British insects that butterflies have clubbed antennae while moths have antennae of many other shapes, but they are never clubbed.

Classification of insects

From the two species studied, the locust and the butterfly, you will have noticed that although they have many characteristics in common, they differ markedly in their life history, their type of mouthparts and their wings. It is differences such as these that are used to classify insects into different orders. The following tables gives details of six orders, selected because they include some of the commonest insects.

TABLE SUMMARISING THE MAIN CHARACTERISTICS ON WHICH DIFFERENT ORDERS OF INSECTS ARE CLASSIFIED

Order	Orthoptera	Odonata	Lepidoptera	Coleoptera	Diptera	Hymenoptera
Examples	locusts grasshoppers	dragonflies	butterflies moths	beetles	flies mosquitoes	ants ichneumons bees hornets wasps
Wings	4 thin and straight	4 transparent and similar	4 covered in scales	4 front pair horny, hind pair membranous	2 hind pair modified as balancing organs	4 front pair hooked to hind pair
Mouthparts	strong jaws for biting	strong jaws for biting	proboscis for for sucking	strong jaws for biting	proboscis for licking (flies), or for piercing and sucking (mosquitoes)	jaws for biting (ants and wasps), or proboscis for sucking (bees)
Life history	3 stages: egg—nymph—imago. Incomplete metamorphosis	3 stages: egg—nymph—imago. Incomplete metamorphosis	4 stages: egg—larva—pupa—imago. Complete metamorphosis	4 stages: egg—larva—pupa—imago. Complete metamorphosis	4 stages: egg—larva—pupa—imago. Complete metamorphosis	4 stages: egg—larva—pupa—imago. Complete metamorphosis

7

Insects: 2

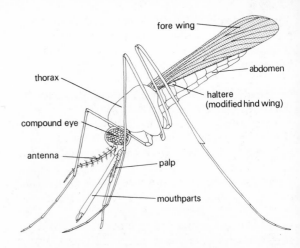

Fig. 7:1 Adult female Anopheles mosquito (appendages shown on one side only).

Insects as carriers of disease

We have already seen with the locust and the large white butterfly that insects can be considerable economic pests. Other pest species which are of great importance include the weevil beetles which get into grain stores, the beetles which attack furniture and the wooden beams of buildings, the termites of the tropics which destroy anything made of wood and even cause houses to collapse, and the clothes moths which make holes in any woollen material.

Insects are also important to man because some of them spread diseases. Diseases are caused by a variety of organisms such as viruses, bacteria, protozoa and worms. They enter the bodies of people, animals and plants and cause various kinds of damage by releasing poisons (toxins) or destroying tissues; in this way they act as parasites. But every parasite must have a means of transference from one organism to another, otherwise the species would die out; insects often act as vehicles for carrying them, and in doing so help to spread the disease. Man has to wage war constantly on some species of insects because many of the diseases they carry are dangerous, e.g. malaria, yellow fever, cholera and plague. Insects transfer these disease-producing organisms in two main ways: by injecting them into the blood stream and by contaminating food.

TRANSFERENCE BY INJECTION

Many insects feed on blood. We have all been bitten by mosquitoes, for example. Let us see what happens when this takes place. The mouthparts are highly efficient tools for the task and include two pairs of needles for piercing the skin (mandibles and maxillae) and a tube for sucking up the blood (labrum). All these lie in the groove of the labium (Fig. 7:2). The latter does not enter the skin itself, but appears to guide the other parts as fingers guide a billiard cue. The tube formed by the

Fig. 7:2 Mouthparts of mosquito in action.

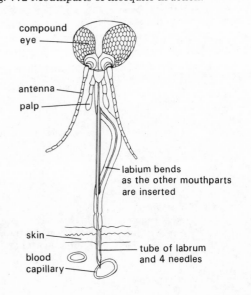

labrum is so narrow that it would quickly become blocked up if the blood coagulated. This is prevented, however, because saliva containing an anti-coagulant is first poured down the tube; the blood can then be sucked up safely. It is the saliva which probably causes the irritation and swelling that follows the bite.

Malaria

It has been estimated that 50 million cases of malaria occur each year, mainly in tropical and sub-tropical countries, and nearly 1 million people die of the disease. The organism which causes it is a protozoan parasite called *Plasmodium*, of which there are several species producing various forms of the disease. Certain species of mosquito belonging to the genus *Anopheles* are entirely responsible for the transmission of the disease. The organisms are injected into the blood stream through the saliva and after a period in the liver cells, they attack the red corpuscles. They multiply inside the corpuscles and burst out again to infect more and more cells. Toxins liberated into the blood when the corpuscles burst produce the characteristic fever of malaria. If a mosquito feeds on the blood of a person suffering from malaria these microscopic parasites are sucked up and undergo a complicated cycle inside the mosquito. As a result, spores are produced which end up in the salivary glands from which they may be injected into another victim who will then contract the disease (Fig. 7:3).

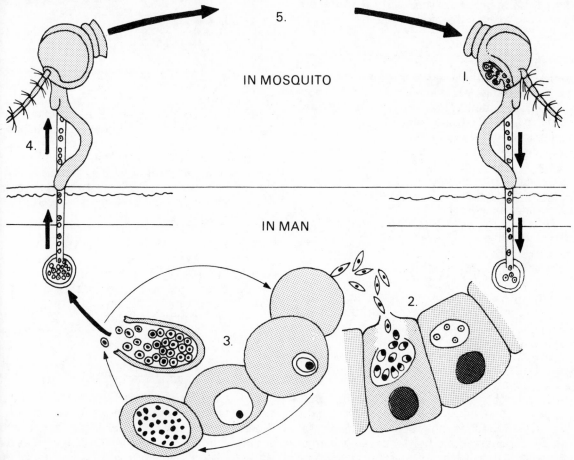

Fig. 7:3 Life cycle of the malarial parasite: 1. Parasites injected into a person's blood system. 2. Multiplication within the liver cells. 3. Red blood cells invaded and multiplication within them (process repeated many times). 4. Parasites absorbed into the body of a biting mosquito. 5. Multiplication within the mosquito's body and passage to its salivary glands.

It is possible to fight malaria·in two ways: by attacking the mosquito which is the essential carrier, or by attacking the parasite itself in the blood. In the former case it is necessary to understand the life history and habits of the species concerned in order to attack the weakest point in its cycle.

Life cycle of a mosquito

A mosquito which is easily obtainable is the common gnat, *Culex pipiens* (Fig. 7:1). Fortunately this species does not carry malaria, but its life history is similar to those that do. Gnats are frequently seen during the warmer months near water. You can find the larval and pupal stages in small patches of standing water such as in water troughs, artificial ponds, in marshes or ditches. To study the whole life history proceed as follows:

Collect a number of females and enclose them in a muslin tent over a bowl of water containing some decaying leaves. The females will then lay their rafts of eggs on the surface of the water. Examine these under a microscope and note their shape and number. The eggs should hatch in two or three days, the larvae forcing their way through the bottom of the eggs into the water.

Examine a larva (Fig. 7:4b) under the microscope in a drop of water on a cavity slide.

First note the head with its feeding brush of hairs, which it uses to waft the micro-organisms on which it feeds towards the mouth.

Now look at the breathing tube which projects from near the end of the abdomen. Inside it you will see two tubes which continue down the body on either side of the gut. These are the large tracheal tubes which supply all the parts with oxygen.

Observe how the gut shows squeezing (peristaltic) movements which help to move the food along inside.

Now watch some larvae in a beaker of water.

How do they take in air through the breathing tube? Deduce from their movements (a) whether the larvae are lighter or heavier than the water; (b) how they remain in the surface tension film.

When they are at the surface, move your hand above them without touching the beaker.

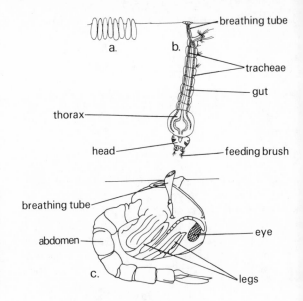

Fig. 7:4 Stages in the life history of the mosquito (*Culex*): a) egg raft b) larva c) pupa.

What happens?

Again, when they are at the surface, tap the bench from below (why below?). What happens?

From your observations you should be able to deduce the stimuli to which they are sensitive and guess the possible usefulness of their reactions in avoiding being eaten by predators, e.g. wading birds.

The larval stage lasts about three weeks, during which there are three ecdyses. The fourth moult reveals the pupa underneath. Unlike the pupae of butterflies they are capable of rapid movement but, like them, they do not feed. They may be distinguished from the larvae by their top-heavy appearance (Fig. 7:4c).

Look at a pupa under the microscope. What adult organs can you see? Look for the two breathing tubes which project like horns from the top of the thorax, penetrating the surface tension film of the water when the pupa is resting. Watch some pupae in a beaker of water and compare their behaviour with that of the larvae. Are they heavier or lighter than the water?

The pupal stage lasts only a few days. The cuticle then splits in the head and thorax region and the mosquito drags itself out, using the old cuticle as a raft while it enlarges and dries its wings before flying off.

> Examine some living mosquitoes in a specimen tube. Note that they only have one pair of wings. For this reason they are classified in the order Diptera. In place of the second pair, look for a pair of small projections called **halteres**. These act as gyroscopic organs, helping the mosquito to keep its balance while flying. What sexes are your specimens? You can tell the difference by the antennae, which are very bushy in the male and much thinner in the female. Note the long straight proboscis. Both sexes can use it for sucking up fluids, but only the female can suck blood; the males feed on nectar from plants.

It can be seen, therefore, that the life history of the mosquito, like the butterfly, exhibits a complete metamorphosis, but it differs from the butterfly in having aquatic larvae and pupae, and in having pupae which are mobile.

Methods of control

In order to eradicate the diseases which are carried by mosquitoes, these insects have to be killed if possible. Here are some of the methods of control which have been used successfully:

1. The larvae and pupae may be attacked by spraying with oil all standing water where they may be breeding. The oil spreads out to form a thin film which reduces the surface tension of the water, so that when they come to the surface to breathe, they sink and water gets into their breathing tubes. This suffocates them. By adding insecticide to the oil this treatment becomes more effective. Vast areas have been sprayed from the air in this way.

There is a particular difficulty, however, in areas where wells are used for drinking water, as the water becomes contaminated with the oil. In parts of India lead-free petrol is used instead, which, after killing the larvae and pupae, quickly evaporates, leaving the water still drinkable.

2. Small fish may be introduced to feed on the larvae and pupae. This method has even been used in parts of India for treating wells. For example, in the Salem district in South India where 11,000 wells serve 300,000 people, the minnow fish, *Gambusia affinis*, has been used. If the wells dry up, they have to be re-stocked.
3. Marshes and swamps may be drained to prevent breeding. This has been effective in many parts of the world and is particularly important in regions within mosquito-flying distance from towns and villages.
4. Houses may be sprayed internally with a persistent insecticide so that if mosquitoes enter and settle they will be killed by contact with it.

By these and other methods malaria and yellow fever have been virtually eradicated from many regions. However, as far as the use of insecticides is concerned, this does not appear to be the final answer because mosquitoes are becoming immune to these poisons. Another problem is that when insecticides are used on a large scale, often indiscriminately, they may have side effects on other forms of wild life, not to mention the human population, sometimes with disastrous results.

Unfortunately, in some places, man has made the malarial situation worse by damming up rivers to produce reservoirs for irrigation and other useful purposes, thus enlarging the breeding grounds of the mosquito.

Yellow fever

Certain species of mosquito carry other tropical fevers. One of the most important is yellow fever. This disease may be present in colonies of monkeys in both Africa and South America.

Fig. 7:5 Spraying houses with insecticide as part of an anti-malarial operation, E. Africa.

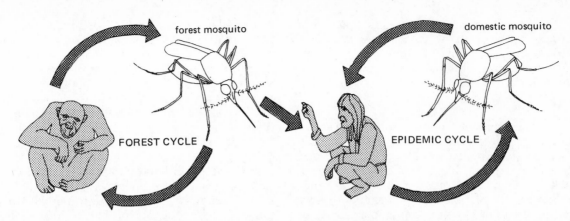

forest mosquito

domestic mosquito

FOREST CYCLE

EPIDEMIC CYCLE

Fig. 7:6 Diagram to show how yellow fever which is carried by a species of forest mosquito from monkey to monkey may affect man and be passed on to others by a species of domestic mosquito, causing a possible epidemic.

It is spread from monkey to monkey by a forest species of mosquito. A person working in the forest might be unlucky enough to be bitten by an infected mosquito and contract the disease. If, however, he travelled back to a town and was bitten by another species of mosquito, the disease might be spread to other people and an epidemic could break out (Fig. 7:6). It is because of this danger that visitors to these areas have to be inoculated against yellow fever before they are allowed to enter or leave the country.

Bubonic plague or black-death

This is a disease which occurs in rodents, especially the black rat. It is carried from rat to rat by means of the rat flea. It is a great danger to humans, because when large numbers of rats die of the disease, the fleas leave the rats and may then bite people. This is what happened in the 14th century when it is estimated that 24 million people in Europe died of the disease. Today it is still present in wild populations of rodents in various parts of the world. It constitutes a potential danger if infection spreads to domestic rodents, as the fleas of these animals are more likely to come into contact with people.

Sleeping sickness

This is a disease which is prevalent in parts of Africa. It is carried by the tsetse fly. The parasite which causes it is a protozoan called a **trypanosome** (Fig. 7:7). The trypanosomes live in the blood stream of wild antelopes and other game, and are transmitted from animal to animal through the bite of the tsetse. If people are bitten by an infected insect they may contract the disease. Some species of trypanosomes spread from the wild animals to domestic species, causing the disease called **nagana**. This makes the keeping of cattle very difficult in some parts of Africa.

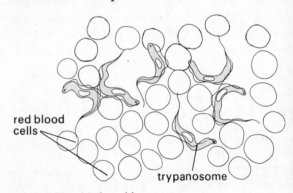

red blood cells

trypanosome

Fig. 7:7 Blood infected by trypanosomes.

Fig. 7:8 Tsetse fly feeding on blood from a finger.

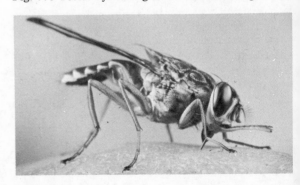

74

SUMMARY OF SOME OF THE DISEASES TRANSMITTED BY INSECTS AND TICKS

Carrier	Parasite causing the disease	Disease
Mosquito	round worms (*Filaria*)	elephantiasis
	virus	yellow fever
	virus	dengue fever
	protozoan (*Plasmodium*)	malaria
	virus	myxomatosis (rabbits)
Fleas	bacterium (a bacillus)	bubonic plague
	virus	myxomatosis (rabbits)
Tsetse fly	protozoan (a trypanosome)	sleeping sickness (man)
		nagana (cattle)
Lice and ticks	bacterium (*Rickettsia*)	typhus
	bacterium (a spirochaete)	relapsing fever

Myxomatosis

This disease, which attacks rabbits, is caused by a virus which is carried from rabbit to rabbit by insects. The disease was introduced into Australia to control the rabbit population which had reached plague proportions. The

Fig. 7:9 Aphids may transfer virus diseases from plant to plant when they feed.

introduction was made in 1950, and by the following year it had spread over an area of $2\frac{1}{2}$ million km², bringing about a very high proportion of deaths. Myxomatosis today is still a major factor in keeping down the rabbit population in Australia, but the rabbits now show much greater resistance to the disease than formerly.

The disease was also introduced into Europe in 1952, reaching Britain in 1953. It spread more slowly than in Australia probably because there it was carried by mosquitoes and in Europe by rabbit fleas. By this means the rabbit population was greatly reduced. Outbreaks of the disease still occur sporadically, but its effect is now more limited.

The table above gives a summary of some of the diseases carried by insects and ticks (which are arachnids).

Plants are not immune from the diseases spread by insects. Many aphids which suck the sap from plants inject viruses into them at the same time and so spread certain diseases; one virus that attacks potato leaves is called potato leaf-roll.

TRANSFERENCE BY CONTAMINATION

The other important means by which diseases are spread by insects is through the contamination of food. The house fly is the most important of these insects, and in order to understand how contamination is brought about you should study a fly's structure and habits.

The house fly, *Musca domestica* (Fig. 7:10)

Prepare two covered Petri dishes, placing a small lump of sugar in one and a drop of sugar solution in the other. Catch some house flies and after starving them for a few hours introduce one into each dish. Observe their feeding methods, using a hand lens or, preferably, a binocular microscope.

House flies can only feed on liquid foods and yet they are attracted to solids. You will see the reason for this if you look at the structure of the proboscis (Fig. 7:11). It is a device for spraying the food with saliva and also for sucking it up. When it dabs at some solid food it sprays saliva over it through fine tubes; the saliva, being a digestive fluid, quickly dissolves the food and the semi-digested liquid is then sucked up by the pumping action of the muscles of the proboscis. Some of this fluid may be forced out again and left behind as a vomit spot. The glass of windows and pictures may become covered with these spots if flies frequently settle on them. You may also have noticed when watching a fly feeding that, as it moves about, it touches the end of its abdomen on the food and deposits a minute drop of waste matter from its gut. Flies also have very hairy legs and bodies to which all sorts of dirt can easily cling.

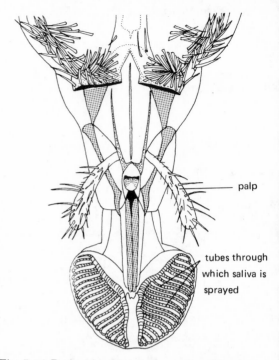

palp

tubes through which saliva is sprayed

Fig. 7:11 Proboscis of house fly much enlarged.

Fig. 7:10 House fly (covering of hairs not shown).

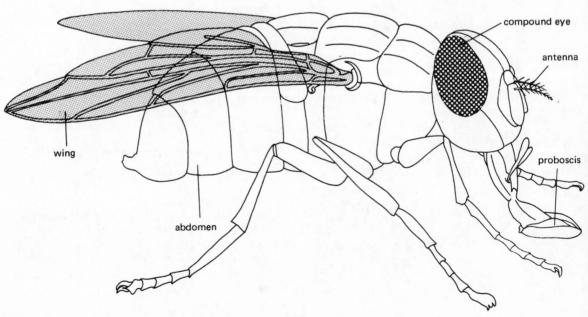

compound eye

antenna

proboscis

wing

abdomen

76

You can now build up a picture of how they carry disease by contaminating food. Unfortunately they visit all kinds of material, including human faeces, and if these are infected with organisms responsible for such intestinal diseases as cholera, dysentery and typhoid, they may transfer them to food. Bacteria may be carried on the hairs of the body or legs, in the droppings, in vomit spots or via the saliva. It is obvious from this that house flies are a menace to man and need to be destroyed if possible. The best way is to prevent them from breeding. Keeping this in mind, we will now study their life history.

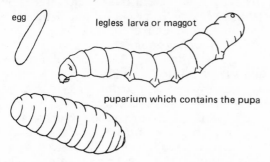

egg

legless larva or maggot

puparium which contains the pupa

Fig. 7:12 House fly: stages in the life history.

Life history (Fig. 7:12)

Eggs are laid in large clusters in decaying material such as manure heaps; they are laid just under the surface so preventing them from drying up. They hatch very quickly (in about 8 hours at 30°C) into almost transparent, white, legless maggots. These larvae feed on the semi-liquid material and burrow into it. After several moults the maggots pupate, but unlike the butterfly and the mosquito the last larval skin is not shed, but remains as a protective coat called a **puparium**. Inside this covering each larva becomes a pupa, from which an adult fly emerges after about three days—a very quick metamorphosis.

Control

From this life history, and from your knowledge of the feeding habits of the adult flies, you will see how important it is to keep flies out of houses and prevent them from contaminating food. To this end food should be kept under covers, and house flies which get into the house should be killed with an insecticide spray. To prevent them from breeding, compost or manure heaps should be as far away as possible from houses; lids should be kept on dustbins which should be emptied frequently and town rubbish should be burnt or sprayed with insecticide. However, some insecticides can be dangerous in other ways, so it is important that the right kind is used. In some towns and cities the house fly has been virtually eliminated by using these methods, but the problem is more difficult in country districts. In some tropical countries the problem is much greater and the risk of disease is far more serious. But once more, a word of warning should be given about the indiscriminate use of insecticides; some are persistent and have dangerous side effects, and their use should be carefully controlled.

USEFUL INSECTS

So far we have studied insects which are mainly harmful to man, either because they are economic pests or because they carry disease, but many insects are useful.

In the tropics, for example, termites have a bad name because of the harm they do to anything made of wood, including houses, but they are also very useful as refuse disposal agents, clearing up the forests of debris and turning the old wood into nutrients which are quickly returned to the soil.

Amongst the most beneficial insects are the many species of bees, of which the honey bee is the most important example. Honey bees are valuable to man, not merely because they produce honey, but mainly because they bring about the pollination of a great many species of flowers without which no seeds or fruits would be formed. We will now study the honey bee in more detail.

The honey bee, *Apis mellifera*

Honey bees are called **social insects** because they live in colonies and have evolved a complex society where there is much division of labour between the individuals. A **colony** in summer contains three kinds of bee: fertile males, called **drones**, of which there are usually several hundred; a single fertile female, the **queen**, which does all the egg-laying and up to 100,000 sterile females, called **workers**, which carry out the work of the hive. In the autumn the drones are excluded from the hive

77

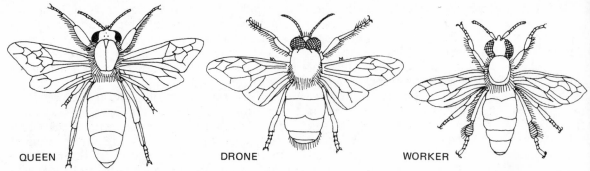

Fig. 7:13 The three kinds of individual in a honey bee colony.

and soon die, and the number of workers is considerably reduced (Fig. 7:13).

The **'nest'** is composed of a series of wax combs which hang vertically and parallel to each other leaving a small gap between each. Each comb is two cells thick; they lie back to back and nearly horizontal. These cells are made of wax which is produced by glands in the abdomen of the workers and modelled by their mandibles. The cells are of three kinds, according to the type of bee which will develop inside. Not all the cells are for breeding; those towards the outside of the nest are used for storing honey and pollen.

Life history

The queen spends her time laying eggs, often as many as 1500 a day. One egg is placed in each cell (Fig. 7:14). If it is placed in a queen or worker cell it is fertilized and develops into a female (queen or worker); those laid in drone

Fig. 7:14 Honey bee: section through cells showing developmental stages.

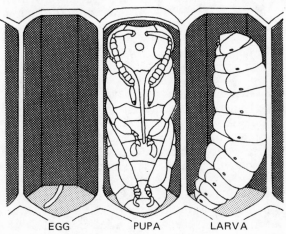

EGG   PUPA   LARVA

cells are not fertilized, but develop nevertheless and become drones. This unusual kind of reproduction when an unfertilized egg develops is called **parthenogenesis**.

The eggs hatch after 3 or 4 days into white legless larvae which are all fed for the next few days on royal jelly, a white milky substance regurgitated by workers from their salivary glands. This is a very rich food. Queen larvae are fed on royal jelly until they pupate, but drone and worker larvae after a few days are given a mixture of honey and pollen called 'bee bread'. It is this difference in the food supplied to queen and worker larvae that causes them to develop differently. The larvae pupate in the cells when 5 or 6 days old after the workers have sealed them in with a capping of wax. The adults emerge by biting their way through the capping after another 10 or 11 days. Thus the colony can grow very fast.

The queen may live as long as 5 years, drones about 5 weeks and workers only 4 weeks in the summer; but workers hatched in the autumn live right through the winter.

Division of labour

Workers perform many tasks which vary according to their age and the needs of the colony. At one time it was thought that they carried out a whole series of tasks in a particular sequence, but this is not so. The life of a worker may be divided into three periods. During the first three or four days the only duty seems to be the cleaning out of the cells before the queen again lays eggs in them. During the second period of about two weeks they wander about the hive a great deal, performing many tasks. These include the feeding of younger larvae on royal jelly; building more comb and capping the cells when ready; regulating the temperature of the hive by clustering over the brood if cold, or fanning with the wings to circulate cooler air if too hot; and receiving nectar from foraging bees and storing it in the cells. During the final period of life, which is usually only about a week or ten days, they forage for nectar and pollen and defend the nest from robber bees or wasps. Thus through division of labour the complex running of the hive is carried out efficiently.

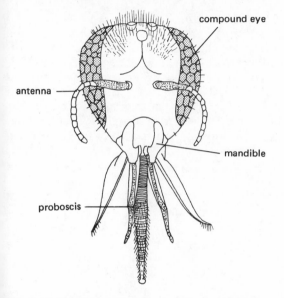

Fig. 7:15 Head of worker bee showing mouthparts.

Food and foraging

Workers collect nectar from the nectaries of flowers. Nectar is a dilute solution of sugar in water which is sucked up with the help of the proboscis (Fig. 7:15). It is carried in the crop, and when the worker returns to the hive it is regurgitated and passed on to other workers. As it passes from worker to worker it is changed by digestive juices into honey which is stored in the cells. When some of the water has evaporated from it, it is sealed off by a capping of wax for later use. Much honey is needed to keep the colony alive during the winter. Pollen is also gathered from the stamens of flowers and stored with the honey; it is rich in protein.

1. Watch bees which are visiting flowers. Notice how each probes into the nectaries with its proboscis, and the way its body becomes dusted with pollen. See if you can make out the way a bee cleans its body with its legs. If you watch by the entrance of a hive (do not stand in front!) you will see that the colour of the pollen varies; this is a clue to the flowers being visited, e.g. apple (grey-green), crocus (orange), horse chestnut (brown).

2. Catch one or two workers in a net, transfer them to a glass-topped box, and examine them under a hand lens or binocular microscope. Notice how they clean their antennae with their front leg.

3. Now remove the legs of a dead bee, examine under the microscope, and compare the structure of the fore, middle and hind leg. Notice the comb on the first leg, which is used for cleaning the antennae, the prong on the second leg for pushing the pack of pollen from the basket, and the 'brush' and 'basket' on the third leg. Can you see how the pollen is held in the 'basket'? Why does it not fall out?

Fig. 7:16 Legs of a worker bee.

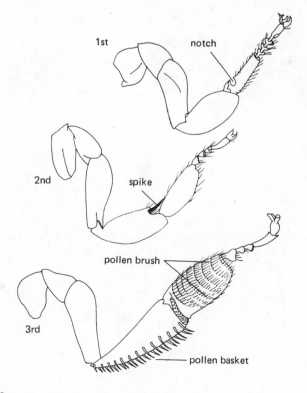

Fig. 7:17 Queen bee surrounded by workers.

Fig. 7:18 Swarm of bees in an apple tree.

Swarming

When a colony becomes large the workers construct queen cells, and when the new queens are nearly ready to emerge, swarming takes place. When this occurs, up to 20,000 workers leave the hive with the old queen. The air is thick with flying bees. The queen soon settles, often on the bough of a tree, and the workers cluster round her until they form a solid mass of bees. They remain in this condition while scout bees look for a suitable place to build a new colony. Sometimes the scouts find an empty hive, a hollow tree or even the chimney of a house, and then the swarm flies off, takes possession of the new home, and

builds a new nest. Meanwhile, in the parent colony, the first new queen to hatch usually kills off her rivals by stinging them while they are still in their cells. If more than one emerges they fight until only one survives. After a few days the successful queen, followed by the drones, leaves the hive for the mating flight. One of the drones mates with the queen while in the air. After separation the queen returns to the hive and starts to lay eggs at once.

Communication

A society works best when there are good means of communication between the individual members. Bees are able to do this largely through their senses of touch and scent. Worker bees learn the lie of the land round their hive like a map; they find the flowers both by scent and sight. What happens if a bee finds a good source of nectar—can it tell others about it? You can find out in this way.

On a sunny day catch a bee in an inverted glass tumbler, preventing its escape with a piece of cardboard. While it buzzes about introduce a drop of sugar solution on to the cardboard. When the bee finds this, it will

Fig. 7:19 Bee-keeper examining a hive.

80

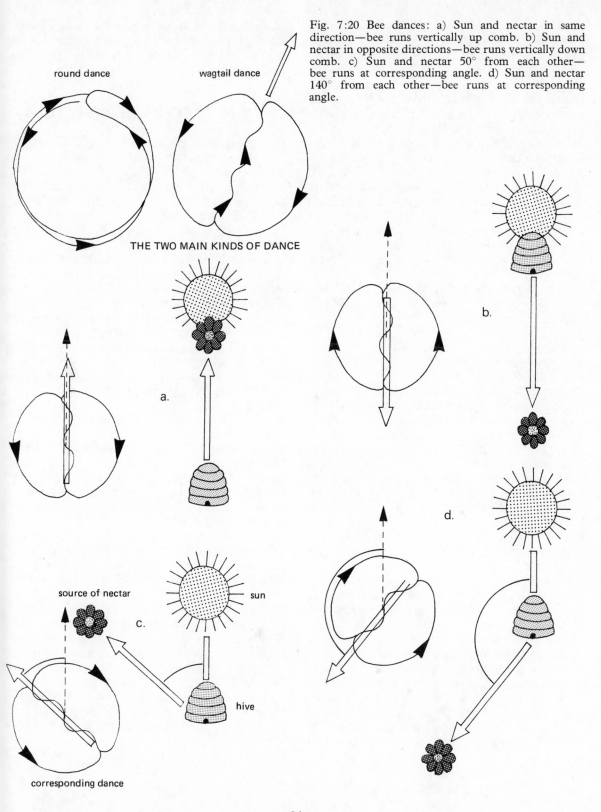

round dance

wagtail dance

Fig. 7:20 Bee dances: a) Sun and nectar in same direction—bee runs vertically up comb. b) Sun and nectar in opposite directions—bee runs vertically down comb. c) Sun and nectar 50° from each other—bee runs at corresponding angle. d) Sun and nectar 140° from each other—bee runs at corresponding angle.

THE TWO MAIN KINDS OF DANCE

a.

b.

c.

d.

source of nectar

sun

hive

corresponding dance

settle quietly to feed. Now remove the tumbler and place a spot of quick-drying cellulose paint on its thorax as a recognition sign; watch what happens. When it flies off, notice its flight pattern. Can you think of a possible reason for this flight pattern? When it has gone, add more sugar solution to the cardboard, which must be left in the same position as before. The bee should return in a few minutes. On finding there is more sugar present, will it tell others about it? Watch to see if others come to the sugar. This would be good circumstantial evidence, but it could be merely chance. What additional experiment could you set up to make sure your conclusion was valid? There are many variations of this kind of experiment that can be done which you might like to follow up, e.g. what happens if you move the cardboard a few yards away while the bee has gone back to the hive. Will it find it when it returns?

The means of communication is by dancing—you can watch this if you have an observation hive, or you can see a film. On returning to the hive after finding a good source of nectar or pollen, a bee will carry out a dance on the vertical comb, the type of dance varying with the distance from the hive. The two extreme kinds are the **round dance** which indicates that the food is only a short distance from the hive, and the **wag-tail dance**, when food is more than 100 m away. In the latter the bee wags its abdomen from side to side as it makes a straight run. Follow the arrows in Fig. 7:20 to see the forms of the dances. In the wag-tail dance the *length* of the straight run varies with the distance of the food from the hive, hence the number of straight runs per minute also varies. For example, if food is 300 m away the worker makes 28 runs per minute, but when 3000 m away, only 9. The *direction* of the food is indicated by the angle of the straight run with the vertical. Running vertically upwards means that the food is in the direction of the sun; running vertically downwards means that it is in the opposite direction.

As the workers crowd round the dancer they touch it with their antennae and thus determine the direction of its run (it is dark in the hive). The dancer also regurgitates a drop of nectar which, together with the scent, indicates the kinds of flowers it has visited.

THE SUCCESS OF INSECTS

Why have they been so successful? Looking back on the species we have studied, certain reasons stand out; you may think of others:

1. *Size*. Their relatively small size means that they need little food to complete their life history and they can live in a great variety of places which may provide both food and shelter from enemies.

2. *Exoskeleton*. This provides them with a very efficient protection against mechanical injury, enemies, disease and loss of water.

3. *Wings*. Flight is an excellent means of escaping from enemies and it enables insects to travel long distances, spread to other habitats and find food.

4. *Excellent sense organs*. These are very varied and sensitive and provide important information about their environment; they include organs of touch, sight including colour perception, smell, taste, hearing and the detection of low frequency vibrations.

5. *Reproductive capacity*. Most insects when they reproduce have very large numbers of progeny. In this way they can multiply very quickly if conditions are suitable. Some, such as bees and aphids, can even reproduce parthenogenetically.

6. *Adaptability*. In the course of evolution they have shown great powers of adaptability, such as in the food they eat (consider their varied mouthparts) and in their many protective devices, e.g. stings and protective coloration.

7. *The ability to live together in colonies*. This only applies to a relatively small number of species, but in these it has been a very important factor in their success. The most important colonial insects are the ants, bees, wasps and termites. All these have evolved complex societies where there is much division of labour between the individuals for the benefit of the whole colony.

8

Fish

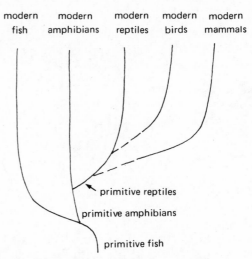

modern   modern   modern   modern   modern
fish   amphibians   reptiles birds   mammals

← primitive reptiles

primitive amphibians

primitive fish

Fig. 8:1 Diagram showing the probable lines of vertebrate evolution.

Adaptation to environment

We have already seen many examples of how the structure of animals is closely related to function. In the next four chapters we are going to study the vertebrates largely from the point of view of how their structure has become adapted to the type of environment in which they are living. The vertebrates illustrate this principle very well, because during the course of evolution there has been a gradual progression from aquatic to terrestrial types. Thus the fish are truly aquatic; the amphibians spend the early part of their lives in water, but as adults live on land; and the reptiles, birds and mammals are typically terrestrial. However, the birds have also become adapted to life in the air, and the mammals, although primarily land animals, have spread to many different habitats, some such as the whales having become adapted once more to the water, while others such as the bats have learnt to fly.

The main lines of vertebrate evolution which have taken place over a period of about 400 million years are indicated in Fig. 8:1.

Water as a medium to live in

Let us first consider the properties of water compared with air.

1. Water is a much denser medium than air. This means two things to a fish living in it: a) it offers considerable resistance to movement, b) it gives support and so reduces the weight to be carried.

2. The temperature of water varies much less than that of air. To the fish this means that it does not have to contend with such extremes of temperature.

3. Far less oxygen dissolves in water than is present in a similar volume of air. For example there are about 8 mg in a litre of water and 250 mg in a litre of air. This is an important factor in respiration.

4. Stimuli are transmitted through water rather differently than through air. For example, light soon becomes absorbed as the depth increases and sound vibrations travel faster in water.

Keep these factors in mind when studying the structural adaptations of fish to life in the water.

The adaptations of fish to life in water

Examine a herring or other similar kind of fish.

Shape
Look at the herring end-on, as if it were swimming towards you. Notice how little of it you can see. Look at it also from the side. Consider both these aspects in terms of resistance to the water when it is swimming; this kind of shape is known as **streamlined**. Ships and submarines follow the same principle.

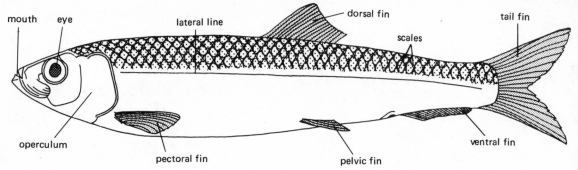

Fig. 8:2 Herring, lateral view.

The mouth

The herring feeds largely on small crustaceans in the water. Correlate this with the size of the mouth. Open the mouth and feel the jaws and note the large number of teeth. How would you expect the teeth to differ in species which feed on other fish?

Operculum and gills

The operculum covers the gills. Carefully cut the operculum away on one side using scissors. You will see the gills underneath; in life these are bright red due to the blood inside them. How many gills are there on each side? Note how each gill consists of a curved supporting

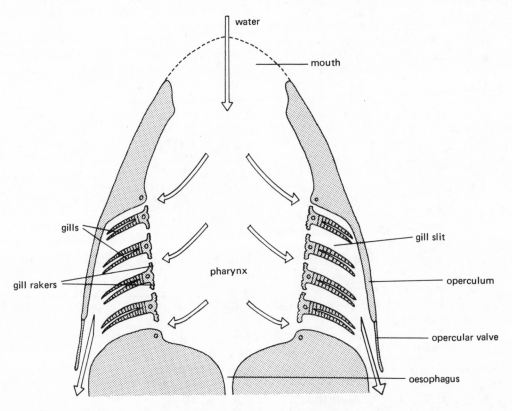

Fig. 8:3 Diagrammatic section through the head of a fish. Breathing takes place in two stages: 1. Water enters when the mouth is opened and the floor of the mouth is lowered, the opercular valve being closed. 2. The mouth is closed, the floor raised, the opercular valve opened and water is forced out.

bar and vast numbers of filaments which contain the blood vessels. Open the mouth wide. Can you make out how the water, which the fish takes in through the mouth, is prevented from passing down the throat into the stomach, but passes instead between the gills and comes out under the operculum? In this process of breathing, the oxygen dissolved in the water diffuses into the blood inside the gill filaments as the water passes over them. Because the total surface area of the filaments is so great this process is very efficient (Fig. 8:3). Now look at the gill rakers on the inside of the gill bar. These act as a sieve which prevents any food passing out through the gills with the water.

Scales

Note how they overlap. Scales are bony plates which protect the fish; they are slippery because of the secretion of mucilage over them. In many fish this mucilage is antiseptic, i.e. it destroys bacteria and fungus spores which settle on it.

Examine some scales under the microscope. Note their concentric growth lines by which the age of the fish may be determined. There is a group of rings for each year of growth, the width varying with the amount of food taken during that period.

Fins

These are supported by fine bony rays, but are nevertheless quite flexible. Look for the pectoral and pelvic fins; they are the only fins which are paired, they are more lateral in position, and they correspond to the fore and hind limbs of other vertebrates. The other fins are single and occur dorsally, ventrally or on the tail.

The tail

This is the part of the body which is posterior to the anus, the opening of the gut in the midventral line. The tail is the main locomotory organ of the body, the tail fin adding greatly to its surface area. Carefully remove the skin from part of the tail region. Note that the underlying muscles are in blocks. When you eat fish the muscles flake off in these blocks. Now make a clean cut through the tail and look at it end-on. You will see that the tail is composed almost entirely of solid muscle

Fig. 8:4 Diagram of the tail region of a herring with the skin removed to show muscle blocks.

supported by the flexible vertebral column. All the muscle fibres run lengthwise down the body.

How does a fish swim?

Now we have seen the structures concerned with locomotion, by observing living fish, we should be able to understand exactly how these structures are used in the process.

Examine any living fish as it swims about in an aquarium. The flexibility of the body depends upon the species; an eel can throw its body into undulations like a snake, but a stickleback cannot. However, all fish move their tails from side to side during locomotion. This movement is brought about by the contraction of the blocks of muscle on one side of the fish in a sequence, starting anteriorly. This is followed by a similar series of contractions on the other side.

When the tail lashes from side to side it produces both a backwards and a lateral thrust against the water. The resistance of the water to this movement causes the fish to go in the opposite direction to these thrusts, i.e. forwards and sideways, for example to the left (Fig. 8:5). When this is repeated from the other side, the fish moves forwards and sideways to the right; the sideways effects cancel out and thus the fish moves forwards in a straight line.

The use of the fins

The tail fin of most fishes helps the efficiency of the tail movements by increasing the surface

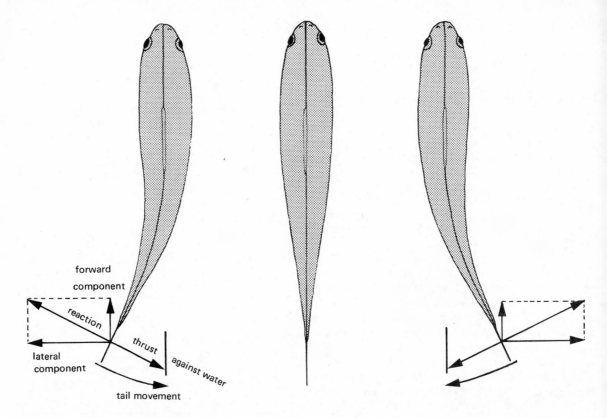

Fig. 8:5 Diagrams showing how the thrust of the tail of a fish against the water results in forward movement.

area and therefore the thrust; the angle at which it is held also has an effect. Note how a fish keeps its dorsal and ventral fins vertical. This increases its lateral surface area and hence the resistance of the water to any sideways movement. In this way these fins act like a keel at the top and bottom, preventing the fish from **rolling** from side to side. They also help to prevent **yawing**. This is the tendency for the body to change its direction from side to side with the thrust of the tail. Now observe how the fish makes use of its paired fins. In most fish, by altering their angle they can cause the fish to swim upwards or downwards in the same way as wing flaps are used in aeroplanes. The pectoral fins are also used as brakes to adjust speed.

The swim bladder

The reason why most fish do not sink in water is that they achieve a similar specific gravity to the water with the help of a swim bladder. This is a long sac-like structure, usually silvery in colour, which lies below the vertebral column in the body cavity. Fish are able to control the amount of gas in the sac and so can adjust their specific gravity according to the depths at which they are swimming. As this adjustment takes time, what do you think would happen if a deep-sea fish was pulled very quickly to the surface on a line?

The sense organs

Most fish have large eyes. Those which live in clear water where light can penetrate easily can see well and many of them can detect colours. What sort of eyes, do you think, would be characteristic of fish which live at great depths?

The nostrils are not used for breathing as in land vertebrates, but only lead to the olfactory organs which enable the fish to 'smell' chemical substances dissolved in the water. This sense is very acute in many fish and enables species such as sharks to find their prey.

86

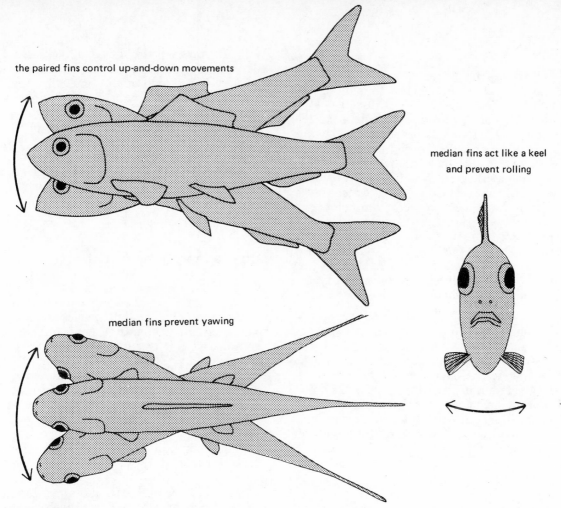

the paired fins control up-and-down movements

median fins act like a keel
and prevent rolling

median fins prevent yawing

Fig. 8:6 Diagrams to show the part played by the fins in controlling a fish's movements.

No ears are visible, but there are internal ears which can detect sound vibrations in the water; these vibrations are transmitted to the ears through the tissues of the body.

Most vibrations of low frequency are detected by the **lateral line** system (Fig. 8:7). This consists of a long canal which runs just under the skin down each side of the fish. There are pores at frequent intervals which connect the canals to the outside, and being filled with fluid, any vibrations in the water outside are transmitted to this fluid and are detected by groups of sensitive cells in the canal. Messages are then sent from these cells to the brain. Thus vibrations caused by the movements of any fish nearby can be detected by the lateral line system; this enables herring to keep in shoals, predatory fish to detect their prey and smaller fish to detect their enemies.

Fig. 8:7 Longitudinal section through part of the lateral line region of a fish.

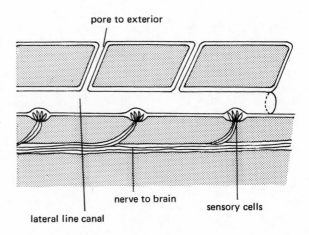

pore to exterior

nerve to brain

sensory cells

lateral line canal

Colour

You saw that the herring was dark above and silvery below; this is characteristic of fish which live near the surface of the sea. How would this colouring help the fish to be less conspicuous to its predators? How well would a shoal of herring blend with its background if viewed by a gull wheeling overhead or by a larger fish looking up from below? What does the disturbed surface of the water in an aquarium look like if viewed from below?

Some fish can change colour in a remarkable manner according to the background. Flatfish such as plaice do this well, and some coral fish can change from one colour to another as they pass by corals of different colours. Try this experiment for yourselves:

Put an equal number of minnows in two glass aquaria which have no sand or gravel at the bottom. Place one on white paper and the other on black so that the paper shows through the glass. Examine the minnows after 24 hours. Have they changed colour? To make it more obvious, now mix the minnows from both aquaria together. Can you distinguish between the two sets?

Reproduction

The herring illustrates how reproduction in fish is adapted to life in water. At spawning time the herring gather in vast numbers in shallower regions of the North Sea. All the females then spawn over a period of a few hours, each laying 20–40,000 eggs which become attached to débris on the sea bottom. At the same time the males liberate the **milt**, a fluid which contains millions of sperms. Thus the water surrounding the eggs is teeming with sperms and the great majority of eggs become fertilized. This is known as **external fertilization** as it takes place outside the body of the female. The young fish soon hatch, as there is little yolk in each egg, and go about in large shoals, feeding on **plankton**, the tiny organisms which live near the surface of the sea.

What do you think are the advantages and disadvantages of this kind of reproduction when so many fish spawn at the same time and so many eggs are laid by each fish? Other fish, such as the salmon, lay fewer and larger eggs; others such as sharks have only two or three, which are fertilized internally, develop inside the mother and hatch, before being born as relatively large fish. Do you think these methods would have any advantages over that of the herring?

Other adaptations of fish

Fish are such interesting animals that you may like to extend your knowledge about them by considering how different species are adapted to different kinds of aquatic environment. For example, would streamlining be more or less important in swift streams? You could check your conclusions by comparing the shape of fish living in fast streams, such as trout and minnow, with those in slow, such as roach, tench or bream. Can you account for any exceptions? How are fish which live on the sea bed adapted to living there? What are the characteristics of those which live at great depths?

You might like to read about the lungfish and how they are able to keep alive when the water in which they are living dries up. How are they adapted to do this?

Why do some fish stay in the same area all their lives while others make great journeys? Find out more about the salmon which spawns in the upper reaches of rivers, but spends most of its life at sea. How do salmon find their way back to the very river in which they were born, when they themselves are ready to spawn?

How is it that eels which are found in the ponds and rivers of Britain and Western Europe never breed there, but migrate right across the Atlantic to the Sargasso sea in order to do so? How do their progeny make the three-year journey back to Europe, and how do they find their way into isolated ponds?

9

Amphibians and reptiles

CLASS AMPHIBIA

From water to land

The class Amphibia is divided into two main groups: those which have tails, the newts and salamanders, and those without, the frogs and toads. These species are present-day representatives of a stage of evolution when some of the vertebrates were beginning to leave the water and starting to colonise the land. Amphibians today can live as adults both on land and in water and show adaptations for both modes of life, but they are still dependent upon water during development.

There is a metamorphosis during their life history. This involves a change from a gill-breathing larval stage, e.g. a tadpole, into a lung-breathing adult, a frog.

It is probably because of the difficulty of being adapted as adults to both land and water that amphibians are not very numerous in the world today, although they flourished and were the dominant vertebrates 250 million years ago when they had fewer enemies. Today the majority of species occur in the tropics where temperature and humidity are high; conditions which call for less extreme adaptations.

As an introduction to this class we will first study an adult frog to see to what extent it is adapted to living on land and in water.

THE COMMON FROG
Rana temporaria

This species is not nearly so common in Britain as it was, because of the draining of many ponds where it bred, and the decline in the number of marshy areas and damp woods which were its favourite habitats. Great numbers have also been killed for scientific and educational purposes; consequently it is important not to kill frogs unless it is essential, and to return them to their habitats when your observations have been made.

First, consider the main problems facing an aquatic animal when it comes on land.
1. It has to support its body during locomotion. Air is far less dense than water hence the weight carried is greater. You can see this for yourself if you let all the water out of a bath before getting out; it takes more strength to get up, and the bath feels much harder to lie on. Thus a land animal has to have much greater support for its body and stronger organs of locomotion.
2. It has to obtain the oxygen needed for respiration from the air. At first sight this should be easier because there is much more oxygen in the air than there is in the water, but there are difficulties. Can you think what they are?
3. It has to cope with wider fluctuations of temperature. On land the temperature ranges from well below freezing to the high temperature in the summer sun, conditions which may lead to death through freezing or through great loss of water from the body.
4. It has to have sense organs which are adapted to receive stimuli coming through the air. Light travels far better through air than through water but sound vibrations travel more slowly. What about chemical stimuli? How do you think the sense organs would be adapted to these factors?

Adaptations for supporting the body and for locomotion

Examine a living frog in a covered glass container or in a vivarium. Unlike the fish, there is no tail for locomotion, and instead of paired fins there are limbs. The limbs are used both for support and locomotion. They are well adapted for both these functions as the bones give rigidity and strength, and the joints and powerful muscles give leverage.

Compare the lengths of the fore and hind limbs and the number of joints in each. Now watch the frog carefully when it hops to see what part is played in the process by the front and hind limbs.

Place the frog in water and watch it swim. Notice the webbing between the toes of the hind limbs. In what way does this webbing increase the efficiency of the action? The great length of the hind limbs makes them more efficient both for leaping and swimming; this is partly due to the elongation of the ankle bones. What is the function of the fore limbs when it swims?

From your observations you will have seen that the limbs are well adapted both for life on land and in the water.

The general arrangement of bones in the limbs is typical of all terrestrial vertebrates, including ourselves. It is said to be **pentadactyl** or five-fingered. A typical pentadactyl limb (Fig. 9:1) is arranged on the 1.2.9.5 plan. Thus there is a single bone in the upper part of the limb, two in the lower, a group of nine (some of which may be fused together) forming the wrist or ankle and five digits (fingers or toes). The number of digits may have become reduced in some species, as it has been in the fore limbs of the frog, which have four.

Respiration

It is possible for oxygen to pass into the blood from the air through any moist surface below which there are thin-walled blood vessels. In the frog there are three such surfaces—the skin, the inside of the mouth and the lungs.

1. *Skin respiration.* Unlike the skin of a fish, a frog's skin has no protective scales, is very thin, and is supplied with many blood capillaries. It is also kept moist when exposed to the air by the secretion of mucilage from glands in the skin, making it feel slippery. When the frog is on land, the oxygen dissolves in this mucilage and diffuses into the blood inside the capillaries while carbon dioxide passes out from the blood by the same method. When the frog is in water the same diffusion process occurs.

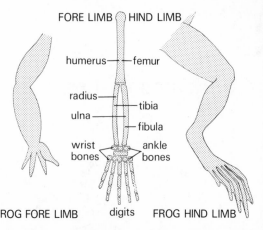

Fig. 9:1 Fore and hind limbs of a frog compared with the bones of a typical pentadactyl limb.

2. *Mouth respiration.* For this to occur air has to be drawn into the mouth cavity. You can see how this is done:

Observe a living frog. Note that the mouth is kept shut all the time. Watch its nostrils and note the rapid movements of the throat. It is muscular action that causes the floor of the mouth to go up and down, thus decreasing or increasing the volume of the mouth cavity. In this way air is forced out or drawn in through the nostrils, allowing the oxygen to diffuse into the blood capillaries under the moist surface of the mouth cavity.

3. *Lung respiration.* Periodically you may see the frog appear to swallow, the body swelling up at the same time. Watch the nostrils when it does this. By shutting the nostrils when the floor of the mouth is raised air is prevented from going out by that route; instead, it is forced into the lungs which swell up considerably as a result. The oxygen then diffuses into the blood capillaries of the walls of the lungs.

Gaseous exchange through the skin occurs all through the frog's life, when on land, in water, and when hibernating, but this method by itself is not sufficient for an active life. On land it uses mouth and lung respiration in addition, the former all the time, the latter particularly after active movement. When the frog is floating in water you may have noticed how its nostrils are above the surface so that it can breathe as if it were on land.

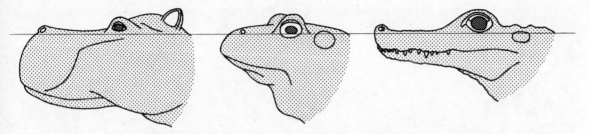

Fig. 9:2 Hippopotamus, frog and crocodile all spend much time in the water but are lung breathers. Note the similarities in the position of the eyes, ears and nostrils.

Sense organs

Chemical stimuli are detected by the olfactory organs which lie at the back of the nostrils. During respiration, chemical substances are carried to them by the air as it passes through the nostrils. These substances dissolve in the film of moisture lining the olfactory organs which become stimulated in consequence. This method contrasts with that of fish where the chemicals are dissolved in the water in which they are living and pass directly in solution to the olfactory organs.

A frog's eyes are very prominent and project above the water when it is floating. Periodically a frog will blink, but a fish cannot do this because it has no movable eyelids. Can you see why the presence of eyelids is a useful adaptation for living on land?

The eardrums, situated behind the eyes, allow the frog to receive sound vibrations both in water and in air. They are tightly-stretched pieces of skin, each being in contact with a bone which transmits the vibrations to the inner ear. The latter is very similar to that of the fish.

The greater efficiency of sight and hearing compared with fish is an important adaptation for living on land, but the position of the eyes, nostrils and eardrums on the head enables the frog to make the best use of these organs when resting almost submerged in the water. This has interesting parallels in other animals. (Fig. 9:2).

Feeding

Frogs feed on worms, beetles, flies and many other insects. The method of capture is adapted to land conditions, sight playing an important part in the process. It is usually when the prey moves that the frog recognises it as potential food. If the prey is slow-moving it will be picked up by the mouth with a sudden jerky movement, but flies are picked off by the tongue. The latter is bi-lobed and covered with sticky saliva and is attached near the front of the mouth. When a frog attacks a fly, the tongue is shot out, its sticky end is wrapped round the insect which in a flash is drawn into the mouth. When it swallows, the eyes are withdrawn into its head, an action which helps to push the food down the throat! (Fig. 9:3).

Fig. 9:3 A frog's method of catching an insect.

91

Fig. 9:4 (top) Several species of fish showing typical external features. (bottom left) Frog fish; what features can you recognise? (bottom right) Flounder, a well camouflaged flat fish.

Life history

Just as in the course of evolution aquatic vertebrates gradually became adapted over millions of years to terrestrial conditions, so during the life history of an amphibian such as the frog we see similar changes as it develops from tadpole to adult. The main stages in the life history are as follows:

1. Hibernation

Amphibians, like fish and reptiles, are 'cold-blooded' or **poikilothermic**. This means they have little control over their temperature which fluctuates according to their surroundings. Thus in countries which have a cold climate they pass the winter in a torpid state known as **hibernation**.

Frogs hibernate in places where extremely low temperatures are avoided such as in the mud at the bottom of ponds or in the banks, below water-level, of streams or lakes. Occasionally a damp situation some distance from standing water is chosen. They usually go into hibernation in October and emerge in February or March according to the weather. They make their way to their breeding places which are usually ponds, lakes or permanent ditches.

2. Mating

Males are usually smaller than females and can be distinguished by their white throats; females have yellow throats. In the breeding season the males also develop a black horny pad at the base of each thumb which is used for gripping the female during pairing (Fig. 9:6). When this takes place the male climbs on to the female's back, the forelimbs being firmly clasped around the female's body, just behind the arms. They move around in this position, sometimes for several days, until egg-laying occurs. The eggs are always laid in water.

As the eggs leave the body of the female the male pours a fluid over them containing sperms; these swim to the eggs and penetrate the thin layer of jelly round each egg before the jelly swells up in the water. Fertilization takes place when the nucleus of a sperm fuses with that of an egg. Thus fertilization is external, but pairing ensures that sperms and eggs come together before the swelling of the jelly makes penetration by the sperms impossible.

Fig. 9:5 Common frog.

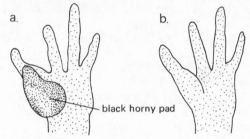

Fig. 9:6 Fore feet of frogs at the breeding season: a) male b) female.

Fig. 9:7 Common frogs pairing prior to spawning, with spawn laid by another frog.

93

3. Development

Development from the time the spawn is laid to the emergence of the small frogs from the water takes about three months at the temperature of the laboratory. Most of the main changes take place during the first three or four weeks after laying, so during this period, especially, make your observations at frequent intervals. Make a diary of the main events and try to relate structural changes to any changes in behaviour you notice. To rear the frogs successfully, proceed as follows:

Place a *small* amount of spawn in any convenient receptacle of water until they hatch. Then transfer the tadpoles into a more permanent aquarium, adding some algae, such as *Spirogyra,* for them to feed on. When they have lost their external gills feed them regularly on finely-chopped raw meat or similar food as by then they are carnivorous. Do not give them more than they can clear up or the water will go bad. When the tadpoles form legs, provide rafts so that the young frogs can leave the water. Release these in a marshy place to fend for themselves.

The following notes will help to emphasise the main points to look out for (Fig. 9:8).

a) *The spawn.* Note how the eggs are spaced out by the swollen jelly. This allows each egg to obtain enough oxygen. Try to pick up the spawn. Its slipperiness will demonstrate how the jelly is an effective protection against pond animals which might try to eat it. The jelly also acts as a cushion, protecting the eggs from mechanical injury.

b) *Early development.* If the spawn has been recently laid the upper surface of the egg will be black and the lower half white, but within a week of laying it will be black all over due to the cells of the black half creeping over and enclosing the white yolky cells of the other half. During this time, as a result of cell division, it will have grown into a ball of cells which looks, under a binocular microscope, rather like a miniature golf ball. The egg then becomes ovoid and later differentiates into head, trunk and tail.

c) *Just after hatching.* Note how at this stage they cling to the old jelly or the side of the aquarium, moving very little unless disturbed. They are clinging by means of a **cement gland** on the under side of the head. They are not eating the jelly as they have no mouth yet, but are still feeding on the yolk inside them which was originally in the white cells of the egg. Two pairs of external gills may be seen.

d) *The external gill stage.* Three pairs of external gills are now present; blood circulates through them and they absorb oxygen from the water. Look at the circulation under the microscope by placing the tadpole in a small Petri dish in a little water and focusing on to the gills.

The tadpoles swim actively at this stage and no longer cling to objects with the cement gland as this is disappearing. The mouth is now functional. Watch how they feed on the algae.

This stage does not last long, so watch carefully for the transition when the external gills shrivel and are replaced by internal ones. The tadpoles look strange during the change over as they have external gills on one side only. This is due to the fact that as the external gills shrivel and the gill slits form, a fold of skin, the **operculum**, grows backwards over them, but it does so unevenly covering the right gills first. Eventually the operculum fuses with the body behind the internal gill slits except for a small opening on the left side called the **spout**.

e) *The typical larval stage.* This is the stage when the tadpole breathes like a fish, taking in water at the mouth and passing it over the internal gills into the opercular cavity and out through the spout. This stage lasts more than two months during which time the tadpole grows considerably. Note how it uses its horny lips to tear the food; also the way the tail moves when it swims. Would you expect the tail muscles to be arranged in the same manner as those of a fish or not?

4. Metamorphosis

This is the change from the typical larva with internal gills to the frog which breathes by means of lungs. The first sign of this change occurs when the tadpoles make frequent trips to the surface to take in gulps of air. This is a

94

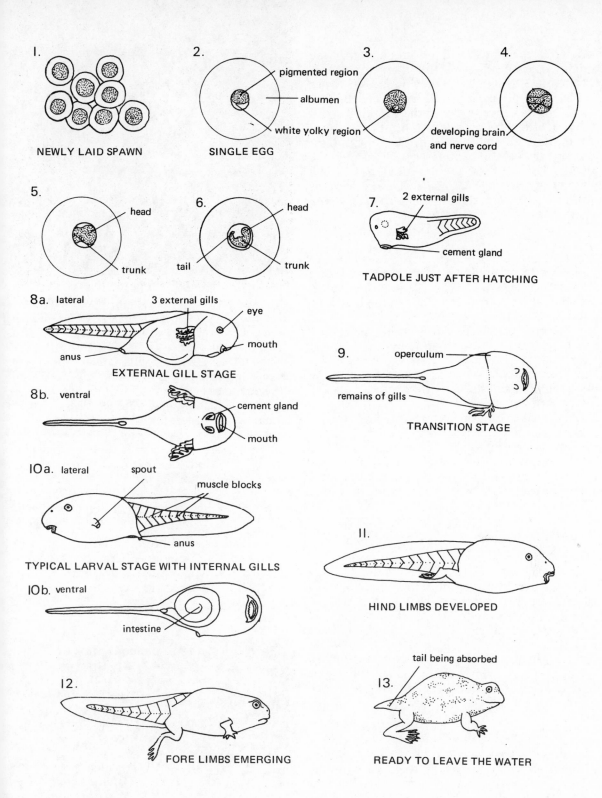

1. NEWLY LAID SPAWN

2. SINGLE EGG
pigmented region
albumen
white yolky region

3. developing brain and nerve cord

4.

5. head
trunk

6. head
tail
trunk

7. 2 external gills
cement gland
TADPOLE JUST AFTER HATCHING

8a. lateral
3 external gills
eye
mouth
anus
EXTERNAL GILL STAGE

8b. ventral
cement gland
mouth

9. operculum
remains of gills
TRANSITION STAGE

10a. lateral
spout
muscle blocks
anus
TYPICAL LARVAL STAGE WITH INTERNAL GILLS

10b. ventral
intestine

11. HIND LIMBS DEVELOPED

12. FORE LIMBS EMERGING

13. tail being absorbed
READY TO LEAVE THE WATER

Fig. 9:8 Frog development.

sign that the lungs are developing. Record the date when you first see this happening. Also watch for the appearance of bumps on either side of the base of the tail. These are the first signs of hind limbs. Later the fore limbs will appear, the left one growing out of the spout and the right one through the operculum. Then the tail becomes absorbed internally, the horny lips are shed, the mouth elongates, the shape of the tadpole becomes more angular and frog-like, and the tiny frogs then leave the water.

Thus metamorphosis involves elaborate changes from a tadpole adapted to life in the water to a frog which is adapted to life on land. The main changes may be summarised as follows:
1. Respiration. A change from gill-breathing to lung-breathing, the skin being used all the time.
2. Locomotion. A change from swimming by means of a tail to walking and leaping by means of limbs.
3. Feeding. A change from aquatic feeding using horny lips which scrape off the food, to terrestrial feeding using tongue and jaws.
4. Perception. The eyes become large and more prominent, and the ears develop and become capable of perceiving sound vibrations in the air.

Toads

These have a very similar structure and life history to that of frogs, but they differ in small details. For example, the skin is not slippery, but is tougher and more warty and it is used far less for respiration. More reliance is put on the lungs which are larger and more efficient than those of the frog. The method of spawning is also different, the eggs being laid in long strings of jelly which are festooned round the waterweed. The details of development are almost identical.

Newts

In many ways newts are much more typical amphibians than frogs as their structure is more intermediate in character between fish and reptiles. For example, they retain their tail when adult and move their body and limbs

Fig. 9:9 Common toad spawning.

in a simpler way. The frog, by contrast, has become highly specialised in structure due to its peculiar method of leaping when on land. This accounts for its short rigid body, the loss of the tail and the elongation of its hind limbs. Keeping these points in mind, it is interesting to keep newts in an aquarium where their whole life history may be observed and their behaviour compared with that of the frog.

Adult newts may be found in ponds and canals between March and June as this is the breeding season. At other times they are more terrestrial in their habits.

Keep the newts in an aquarium, providing resting places for the adults above the surface of the water. Take precautions to prevent them from climbing out. Feed them on small live worms. Make a diary of the events you observe. Here are some points to look out for:

Fig. 9:10 Palmate newts during courtship: the male (left) is vibrating its tail.

Fig. 9:11 a) Smooth newt larva. b) Fringe-toed lizard from S. Europe. c) Slow worm—a legless lizard. d) Adult crocodiles. e) Crocodile eggs removed from nest in the process of hatching. f) 2-day old crocodiles. g) Venom being extracted from a puff adder.

1. The colour differences between the sexes at the breeding season, and the presence of the crest in the male.

2. Courtship. When this occurs, the male tries to attract the attention of the female by displaying its colours and crest and by vibrating the tail which is curled round parallel to the body. The male then lays a minute jelly-like packet of sperms which the female picks up by means of the **cloaca** (a ventral pouch which receives the openings of the food and reproductive canals). Inside the female's body the jelly dissolves and the sperms are freed and fertilize the eggs internally before they are laid.

3. Egg-laying. To do this the female clambers over water weed and lays the eggs singly on a leaf; the leaf is then bent over so that the egg lies between the two parts of the leaf like the inside of a sandwich. Look out for leaves which are bent over in this way.

4. The larval stage. Note how fish-like the tiny transparent larvae are when they hatch. They resemble frog tadpoles by having feathery external gills. Supply them with small crustaceans such as *Daphnia* and watch how they feed. They are carnivorous from the start.

5. Metamorphosis. Note how old they are when they start to grow their legs. Which pair develops first? When do they start coming to the surface to take in gulps of air? When do they lose their external gills? When do they first leave the water?

CLASS REPTILIA

This class includes the snakes and lizards, crocodiles and alligators, turtles and tortoises. Reptiles may be considered true land animals as they can be independent of water as a medium in which to live, in spite of the fact that some, such as turtles and crocodiles, spend much of their life in water. Their main adaptations for living on land are:

1. The lungs have enlarged sufficiently to supply enough oxygen for all their needs, hence the skin is no longer needed as a respiratory surface. This allows the skin to have a hard protective covering of epidermal scales, i.e. scales formed from the outer layer. This is characteristic of all reptiles.

2. Reptiles have solved the problem of having an aquatic larval stage and being dependent upon water by laying relatively large eggs with protective leathery shells which reduce the water loss. As the embryo develops, membranes are formed which completely enclose it in a bath of fluid which acts like a minute pond of its own. Birds and mammals also use similar membranes.

Reptiles are of great interest, but it is only possible in this book to indicate briefly some of their remarkable adaptations.

Body temperature

Like amphibians they are poikilothermic, their temperature changing with that of their surroundings. When on land they are subject to wide fluctuations, and these affect the speed of their physiological processes. Thus when they become hot these processes are speeded up and they are able to be very active, when cold they become sluggish, and when very cold, quite torpid. Hence those which live in places where winters are cold are forced to hibernate.

Reptiles have many adaptations which enable them to avoid extreme temperatures. Desert reptiles, for example, avoid great heat during the day by retiring to a burrow, burying themselves in the sand or sheltering under a boulder. They do most of their feeding at night when it is cooler.

Crocodiles are able to keep a remarkably constant temperature by keeping to a regular pattern of behaviour. At night they are mainly aquatic and are warmed by the water; before dawn they haul themselves out and then bask in the morning sun. As the temperature rises they cool themselves by opening their mouths; this allows the water to evaporate from the mouth surface, taking heat from the body as a result. During the hottest part of the day they often retire to the shelter of trees, but return to bask in the sun in the late afternoon. By behaving in this way, their temperature only fluctuates a few degrees during a 24-hour period.

Feeding

Reptiles have evolved some very unusual feeding adaptations. Chameleons have developed a highly specialised tongue which can be shot out like a released spring by muscular action. This enables them to pick off an insect as far away as 10 cm.

Snakes have specially adapted jaws which allow them to swallow eggs or prey much larger in diameter than themselves. To do this the two bones comprising the lower jaw are freely-movable units attached together in front only by an extensible ligament, and at the back they are not rigidly jointed to the skull but are connected to it by another bone which acts like an extensible hinge. This device allows large objects to be swallowed. The rows of backwardly-projecting teeth also help the swallowing process.

Some species of snakes such as boas and pythons twist their bodies round their prey and suffocate them by the pressure of their coils before swallowing them. Others, such as adders and cobras, immobilise their victims with poison using certain teeth which have become modified as poison fangs. In the adder the two fangs are hollow like hypodermic needles, and poison from glands in the head is squirted through them into the wound made when it strikes the prey. One advantage of feeding on relatively large prey is that the snake does not need to feed as frequently as other reptiles.

Locomotion

Methods of locomotion vary greatly. The most basic method is that used by lizards. This is very similar to the walking gait of the newt, but is quicker, and the body is lifted higher off the ground.

Snakes and certain lizards such as slow-worms have lost their legs during the course of evolution, but by the undulations of the body and the gripping power of the overlapping scales on the ventral side, they are able to move fast. This legless condition also enables snakes to burrow into the ground and thus find shelter. The flexibility of the snake's body also allows many species to climb very effectively; in this way they can reach the nests of birds and feed on their eggs.

Turtles have limbs modified as flippers for swimming and they are so well adapted to an aquatic life that some species only come on land in order to breed.

Reproduction

Reptiles all have internal fertilization. Usually they lay eggs. The eggs are relatively large because they contain much yolk; they also have tough, leathery shells. When the young hatch they are large enough to move actively and fend for themselves.

Successful incubation of the eggs is dependent upon their receiving both warmth and moisture. Various devices are used to keep the conditions suitable. Grass snakes often lay their eggs in heaps of rotting vegetation where extra heat is provided by bacterial action during decay. Turtles come ashore and lay their eggs in the sand at a depth where the eggs receive enough heat, but not too much, from the sun. Crocodiles select sandy situations near rivers and time their egg-laying to coincide with the onset of the dry season, thus avoiding the risk of flooding.

Crocodiles also show parental care. For three months the female zealously guards the nest against such marauders as monitor lizards and marabou storks. When ready to hatch, the young make loud croaking noises which act as a signal for the mother to scoop away some of the sand so that the young can emerge successfully. The mother also protects the young for some weeks after hatching.

Some reptiles (e.g. adders and some species of lizards) do not lay eggs, but retain them in their bodies until they hatch. It is usual for them to bask in the sun, the extra heat helping to incubate the eggs inside them. The young are born as active little creatures quite capable of fending for themselves.

10

Birds

Fig. 10:1 Whinchat alighting.

Life in the air

Birds are terrestrial animals which have taken to a largely aerial existence. Hence they are basically adapted for life on land, but many of their characteristic features are adaptations for flight. Let us consider the main characteristics of birds and see to what extent they are related to life in the air.

Homoiothermy or warm-bloodedness is a characteristic they share with mammals. It is the ability to keep the temperature of the body fairly constant in spite of fluctuations in the environment. Homoiothermy is useful both for a terrestrial and an aerial existence and it enables birds and mammals to live in both extremely cold and hot places.

The structural characteristics of birds are best examined on a dead bird and by looking at its skeleton, but it is also essential to watch living birds to see how these structures are used.

Examine any dead bird; the example we shall use is a pigeon.

Shape

Notice the general streamlining of the body and the way all the feathers overlap in a particular direction to keep the surface smooth. Only the legs disturb the streamlined effect. How do birds overcome this disadvantage during flight? Try to think of any similarities of design between birds and aeroplanes as you study their structure.

Legs

Stretch out one of the legs, measure it and compare its length with the length of the body, that is from the base of the beak to the base of the tail. The ratio varies according to the habits of the species: wading birds have particularly long legs, swifts have very short ones. Long legs enable a bird to take off without hitting the wings on the ground with the first downstroke. How do swifts manage? Note how the lower part of the leg and the toes are covered with scales which are very similar to those of reptiles. Birds usually have four toes, but they differ considerably in position and structure according to their habits. Pigeons are typical of perching birds in having three toes in front and one behind; this arrangement enables them to grip a branch very tightly. Bend the leg as if the bird was squatting, note what happens to the toes. This should help you to explain why a bird does not fall off its perch when it goes to sleep. Look at Fig. 10:2 for other adaptations of the feet.

Beak

This is a projection of the bones of the skull and consists of two mandibles covered with a horny material. The shape varies considerably according to the typical food of the species. You will see other examples in Fig. 10:4. By looking at the beak of any species you come across you should be able to guess the type of food it eats. Try this and check from a bird book whether you are correct.

Sense organs

Examine the large eyes of the pigeon. Find the third eyelid which lies inside the other two. Pull it with forceps and note how it can sweep right across the eye. You will see the nostrils

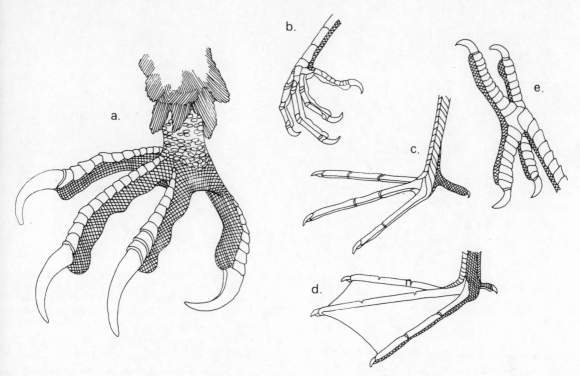

Fig. 10:2 Adaptations of birds' feet. a) Grasping and killing prey e.g. eagle. b) Perching e.g. starling. c) Walking on mud e.g. heron. d) Swimming e.g. duck. e) Gripping on to bark e.g. woodpecker.

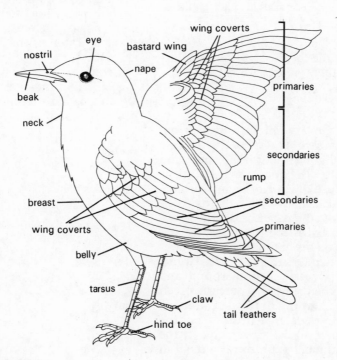

Fig. 10.3 External features of a bird.

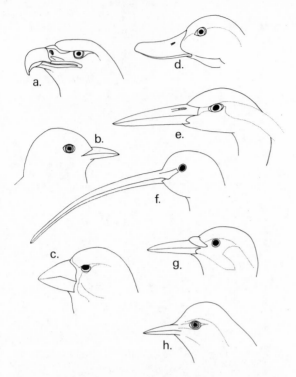

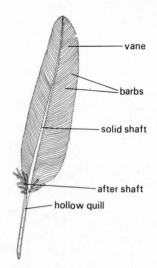

Fig. 10:5 Flight feather.

Fig. 10:4 Adaptations of birds' beaks for feeding. a) Flesh e.g. eagle. b) Varied diet e.g. blackbird. c) Seeds e.g. finch. d) Water plants and animals e.g. duck. e) Fish e.g. heron. f) Animals in mud e.g. curlew. g) Insects in wood e.g. woodpecker. h) Insects e.g. warbler.

opening out of the beak, but the ears are covered with feathers; search for them behind and below the eyes.

Feathers

Examine one of the large flight feathers from the wing and compare its structure with Fig. 10:5. Notice how light it is, but at the same time how strong. Note also how the barbs cling together to produce a wind-resistant surface; how much of a pull does it take before they separate? Now see if you can repair the break by stroking the barbs between finger and thumb from the rachis end outwards. This is what happens when a bird preens its ruffled feathers. First it twists its neck round and probes near the base of its tail, where the preen gland is situated, to obtain a little oil on its beak; next it proceeds to spread it on the feathers; then with special movements of the beak it smooths the barbs until they are all in place. Oiling keeps the feathers waterproof and

prevents them from becoming brittle. Dip a flight feather momentarily in water and notice if it is water-resistant.

To understand the means by which the barbs cling together, cut out a small portion of the vane, separate one or two of the barbs by pulling gently, and examine under the microscope. Notice how each barb has barbules coming from it on both sides; on one side the barbules bear hooks and on the other there is some device such as a ridge which causes the hooks to catch as they slide across each other (Fig. 10:6).

Look for other kinds of feathers on the pigeon. The most obvious are the **coverts** which look like miniature flight feathers. They cover the body all over and fill in the gaps

Fig. 10:6 Diagram of part of a flight feather much enlarged.

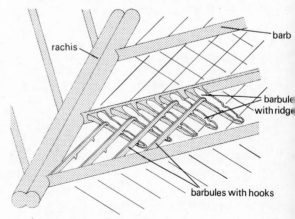

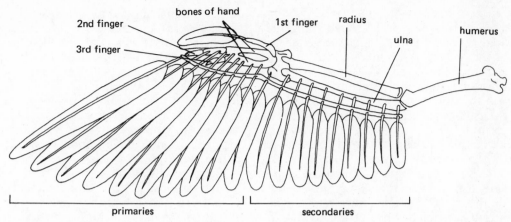

Fig. 10.7 Skeleton of a bird's wing showing the arrangement of the flight feathers (coverts removed).

between the quills of the flight feathers to make a completely wind resistant surface. Underneath the coverts you will find the down feathers which are fluffy because their barbules have no hooks and thus do not stick together. The air trapped next to the body by these feathers acts as an insulator, preventing loss of heat from the body.

Wings

These are modified fore limbs which have become adapted for flight. Stretch out one of the wings and note first its general shape. Compare the thickness of the leading edge with the trailing edge. How does this shape compare with the wing of an aeroplane in section? Notice how the under-surface of the wing is slightly concave. How would this make the downstroke more effective?

The wings of different species vary greatly in shape. Is the shape in any way connected with speed of flight or rapidity of wing beat? You could come to some conclusions about this by comparing the silhouettes of fast fliers such as swifts and hawks, heavy birds such as herons and swans and those which glide a lot such as gulls and eagles. Are there any similarities with the shapes of aircraft wings? Consider especially those aircraft which are built primarily for speed, for carrying heavy loads or as gliders.

The large surface area of the wings is partly due to the feathers and partly to the elongation of the limb bones. In some birds the wing span is enormous, e.g. up to 4 m in the wandering albatross.

Skeleton and muscles

Examine the skeleton of a wing (Fig. 10:7). The arrangement of the bones is not in a perfect pentadactyl plan although it has been adapted from it. Not all five fingers are present and two of the hand bones have fused with those of the wrist. The largest flight feathers, the **primaries**, are attached to the wrist and hand region, while the **secondaries** are attached to the ulna.

Fig. 10:8 Skeleton of a pigeon.

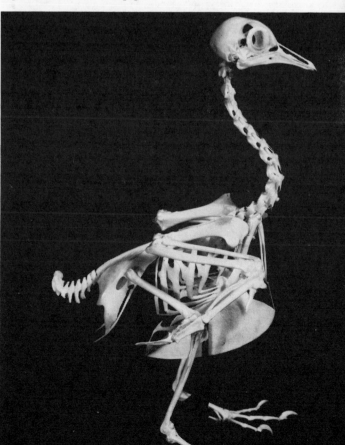

103

Look at the general form of a bird's skeleton (Fig. 10:8) and see how it is related to flight. Compared with a mammal skeleton, such as our own, the body is very rigid due to the fusion of some of the parts. This is because the great flight muscles have to be attached to something very rigid to bring about effective movement of the wings. These flight muscles are attached to the sternum which has a deep ridge down it like a keel. The muscles fit snugly on either side of it. Lie the pigeon on its back and feel the keel in the mid-line and the big muscles on either side. Pluck the feathers from this region and carefully take off the skin. Now remove the flight muscles completely by cutting flush with the keel and the base of the sternum. Note that there are two large muscles on each side which make up this mass: the **major pectoral** is the larger and the **minor pectoral** lies beneath it. They are both firmly attached, with one end fused to the sternum and the other to the humerus. When the major pectoral contracts it forces the wing downwards and slightly backwards, then the minor pectoral contracts and brings the wing up again. The difference in action is due to the position of insertion on the humerus of the tendons of the two muscles, the major acting from below and the minor from above. Look at Fig. 10:9 to see how this happens.

You might think that the upstroke during flight would counteract the lift given to the downstroke, but this is not so, because in the downstroke the flight feathers overlap and lie flat and no air passes between them, but in the upstroke their angle is altered so that the air does pass through. During flight the angle of the flight feathers alters during the downstroke, the primaries providing most of the backthrust and the secondaries mainly the upthrust.

Weigh the flight muscles you have removed from both sides of the sternum and compare with the weight of the whole bird. In most species these muscles are about one-sixth of the total weight, in fast fliers, like pigeons, even more. Effective flight in both birds and aeroplanes depends upon the ratio between power and weight; the higher this ratio, the more effective the flight. It is because the flight muscles produce the power when they contract that they make up such a high proportion of the total weight. The other adaptation towards effective flight is the reduction in weight of the skeleton; birds achieve this by having hollow bones.

The rigidity of the body, which is so good for flying, brings disadvantages in manoeuvrability. This is offset by the great flexibility of the neck made possible by the large number of

Fig. 10:9 Diagram showing how the flight muscles of a bird are attached.

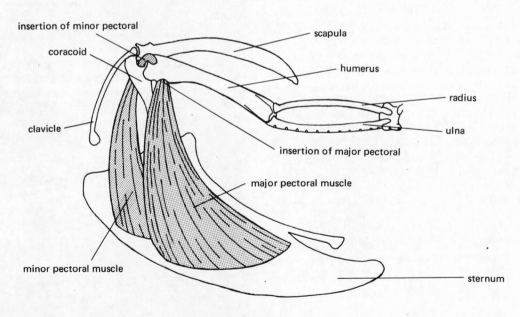

neck vertebrae. Even a short-necked bird like an owl can twist its neck through an angle of more than 180°! Observe how all species of birds, whether perched or flying, are constantly moving their heads about while feeding, preening or watching out for enemies, but their bodies remain rigid all the time.

Breathing

It is largely because of the efficiency of their breathing system that birds do not get out of breath. If you see one panting it is because it is too hot; by passing cooler air in and out rapidly, it cools itself.

Birds have lungs but the breathing method is different from that of mammals (Fig. 10:10). When the bird's sternum is lowered air is sucked down the windpipe and passes right *through* the lungs into a series of air sacs which act as reservoirs; when the sternum is raised the air passes out of the air sacs and through a series of minute tubes in the substance of the lungs. Oxygen is extracted by blood capillaries which surround the tubes. In flight the movements of the sternum correspond to the wing beats, thus the faster it flies the more oxygen it can absorb. This is an excellent method because no stale air is left in the lungs.

We can now summarise the main adaptations of birds to life in the air:
1. The fore limbs are modified as wings.
2. Feathers are present which are light, strong and flexible and which provide a large surface area.
3. The shape is streamlined.
4. The skeleton is extremely light, the bones being hollow.
5. The body is rigid, thus providing a firm base for the attachment of the flight muscles. The sternum is keeled to aid this attachment.
6. The neck is long and flexible. This compensates for the rigidity of the body and allows free movement of the head.
7. The legs are long, thus helping take-off.
8. Eyesight is excellent. This makes high-speed aerobatics and precise landings possible.
9. The efficient respiratory system and the homoiothermic condition both help to increase the efficiency of flight.

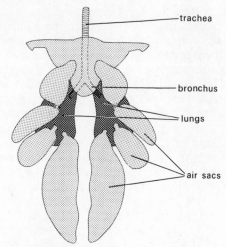

Fig. 10:10 Respiratory system of a bird.

Reproduction

All birds have internal fertilization and lay a small number of relatively large eggs (Fig. 10:11). Each has a hard protective shell which is calcareous and permeable to air (why?). They are laid in a nest which helps to protect them, and are kept at a temperature usually higher than that of the air. This is achieved by a process of incubation when the parent keeps them close to its warm body, usually by sitting on them. Incubation also helps to prevent too much evaporation of water from the egg. In an artificial incubator for rearing chicks air has to be kept very damp for the same reason; otherwise they will die.

During development the chick embryo inside the egg becomes covered with a membrane, the **amnion**, which protects it and encloses it in a bath of water. Blood vessels spread out over the yolk surface and the blood transports the food in solution from the yolk to the embryo. As the chick grows the yolk sac becomes smaller and by the time it

Fig. 10:11 The structure of a hen's egg.

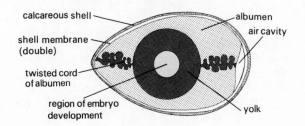

hatches (after 21 days) almost all this food is used up and the chick is able to feed through its beak. When a chick hatches it is covered with down feathers and can move about almost at once, but in most species of birds the young are naked and helpless. However, they soon acquire feathers and after a few weeks are able to fly. During the period at the nest they are fed by the parents and grow very rapidly.

Looking after the young until they are independent is known as **parental care**; it is a very important factor for survival. Compare this with what happens in the majority of fish, amphibians and reptiles, where the young have to fend for themselves after hatching. Can you see any connection between parental care and the laying of only a few eggs?

Observe the behaviour of a common species at the breeding season. By putting up nest boxes some months before nesting time, it is likely that they will be occupied and this makes observations easier. Here are some of the things to look out for:

Note when nesting commences, what materials are used, when egg-laying starts, when the full clutch is laid, whether incubation is carried out by one parent or both, how often the young are fed, what food is brought, how the nest is kept clean, when the young fly and whether they keep with their parents or not. It is essential to disturb the birds as little as possible, or they may desert their nest.

Breeding adaptations

Different species show different kinds of adaptations which make reproduction more effective. See if you can arrive at some of the principles involved by considering these questions:

Why should an owl's eggs be round and a guillemot's pear-shaped? Has the shape anything to do with the nesting habits of these species?

Why should eggs differ so much in colour? Why should birds such as woodpeckers and kingfishers which nest in holes or tunnels have white eggs, while those of ground-nesters, such as plovers and terns, are mottled?

Why should the plumage of birds differ so markedly between the sexes in some species such as ducks and pheasants while others such as kingfishers show no differences?

Territorial behaviour

Most birds have **territories**. These are areas which are defended against other members of the same species. They may be defended by a male, a female, a pair or even a group of birds, for the whole year, or only during the breeding season. The main function of a territorial system is to ensure that the birds of one species are spaced out so that each can obtain enough food for itself and for bringing up its young. There is usually a correlation between the size of the territory and the availability of the right kind of food within it for that species. Thus some birds, such as robins and warblers, have territories as small as 2000 m^2, while eagles may dominate an area of more than 70 km^2.

Birds which nest colonially such as gannets and gulls have individual territories not much larger than the pecking distance between the nest owners, but these territories are not concerned with food supplies as food is obtained by ranging over considerable distances. The great advantage of nesting together is that the entire colony can act together to defend the whole breeding ground.

Territories are defended and the boundaries between adjacent territories are determined by a variety of threat postures and sounds. You can observe this for yourself very easily by watching robins. You will notice how a robin will sing from a certain tree or bush. The song serves to warn other robins that the territory is occupied and will be defended. Singing is an important aspect of aggressive behaviour; birds sing *at* each other, just as some people when they quarrel, shout at each other.

Boundaries between territories are at first vague, but they become more exact as a result of border clashes. If robin A trespasses on to B's territory, B will try to chase it off. If A holds its ground, B will fly up to it and display its red breast with a threatening posture. Usually this is enough to cause A to fly back to its territory, but if this does not happen a fight will develop. Fights to the death, however, are extremely rare. If, on the other hand, B trespasses on to A's territory, it is A which shows most aggression and hence B retreats. It appears that a bird's degree of aggression becomes less the further it strays from its own territory, thus the boundary is the region where both birds show equal levels of aggression. By first noting the places from which the neighbouring birds sing and then watching where scuffles and displays take place, it is possible to map out the territories.

Bird movements

Birds are said to be **resident** if they occur all the year in one area. In places where the seasons of the year are markedly different, the resident birds have to be very adaptable to the changing conditions. Often they have to change their feeding habits. Some species, however, make use of their power of flight to leave the region where they have bred and **migrate** to other parts where conditions are more suitable.

Migrants such as swallows, martins, fly-catchers, nightjars and many species of warbler are insectivorous. They breed in Europe where there is plenty of food for them in the summer, but migrate to Africa in the autumn where an abundance of insects is available. Others such as redwings, fieldfares, bramblings and many species of duck and geese breed during the short summer in the northern tundra, but spend the winter in Britain and the more southern regions of Europe where the climate is mild enough for them to find sufficient food.

The most spectacular migrant is the Arctic tern which breeds within the Arctic circle and then migrates over some 16,000 km of ocean to spend the southern summer in the Antarctic.

Much of our knowledge about migration has been gained from the practice of **ringing**. Light metal rings, suitably numbered, are placed on the legs of fledglings and adults, and details of age, sex, date and locality are recorded for each ring used. The birds are then released. These data are then sent to the headquarters of the ringing scheme for reference, in the event of the ring being reported subsequently. Many thousands of birds have been ringed in different countries and the information has been used to map out the migratory routes. Additional information has been obtained by watching migration with the help of radar, but this is only successful for relatively short distances.

How do birds navigate over these vast distances?

It seems almost miraculous that a swallow can find its way back to the same barn where it nested the previous year after spending the summer in South Africa, or that a young martin or cuckoo can find its way from Europe

Fig. 10:12 The migration routes of the Arctic tern and swallow.

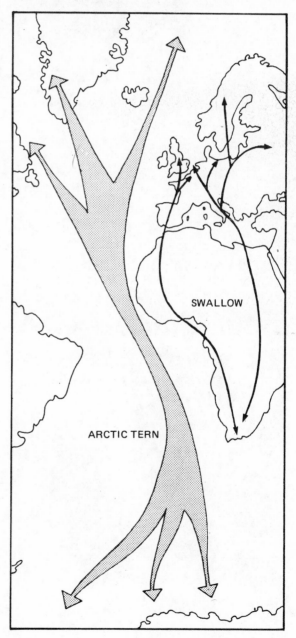

Fig. 10:13 Representatives of a few of the main bird groups: a) Flightless bird —ostrich. b) Waders—ruff (left), stilt (right). c) Birds of prey—martial eagle. d) Swallows and martins—European swallow. e) Woodpeckers—green woodpecker. f) Finches and buntings—goldfinch. g) Tits—great tit. h) Ducks and geese—mallard.

10.14 a) The inconspicuous nest and eggs of the little tern. b) Hen mallard on nest: another example of camouflage. c) Male blackcap being ringed.

109

to Africa, which it has never visited before, without any guidance from adult birds.

There is still much which is unknown, but there are important clues towards an understanding of how some species find their way.

Some birds orientate by the sun. This has been proved experimentally by keeping birds in large cages and altering the apparent position of the sun by means of mirrors. The birds adjusted their position according to the angle of movement of the mirror. It was also found that if the sky was totally obscured there was no orientation at all. What was more astonishing was that they were able to compensate for the natural 'movements' of the sun during the day, because if kept under an artificial sun they orientated at different angles to it according to the time of day; in other words they treated the artificial sun as if it were moving!

Many species migrate at night and orientate according to the pattern of stars. This has been demonstrated by putting some migrants in a planetarium. When the star pattern in the dome was rotated, the birds changed their position in a corresponding manner!

You can now see why under heavy overcast conditions migrants may lose their way. They may also be blown hundreds of miles off course and perish in vast numbers.

Migration is a fascinating subject to read more about.

Projects on birds

Here are some suggestions:

1. Try to find out which of the three main senses a bird uses most. For example, can they discover strong-smelling food which is covered up? Can they see colours? When a blackbird or thrush searches a lawn for worms does it see them, hear them, or smell them?

2. Erect a bird table in a quiet place where it can be viewed from a window. Find the preference foods of the species which visit it by putting out a selection of different kinds.

Make notes on the feeding techniques of the various species. Note any aggressive behaviour between different species. Are some more timid than others? How does their behaviour affect the amount of food each species obtains? Is any particular species earliest to start feeding in the morning or last to finish in the evening?

3. Make a bird bath by digging a shallow hole and lining it with polythene sheeting. Place stones around the edge to anchor the sheeting firmly. Note how the various species that visit it use their beaks when drinking. Note also their bathing techniques. Are their bathing habits correlated with feeding times, temperature or other factors?

4. How fast do birds fly? Time different birds with a stop-watch as they pass between two fixed points such as between two trees or two hedges. Then measure the distance flown and calculate the speed. How would you allow for the effect of wind?

5. Visit different habitats such as a housing estate, wood, marsh or sea shore and identify with the help of a bird book the birds which are characteristic of each. Look out for any adaptations which help them to live successfully in these particular habitats.

11

Mammals

Fig. 11:1 The European badger.

Spreading to all habitats

Mammals represent the peak of vertebrate evolution, although the number of species living today is only about 5,000. They range in size from a species of shrew about 5 cm long and weighing 2·8 g to the mighty blue whale of up to 130 tonnes.

Like the birds, mammals are homoiothermic. This ability to keep their temperature reasonably constant irrespective of their surroundings has enabled them to spread to the coldest and hottest places of the earth. They have also shown great powers of adaptation to varying habitats and different ways of life.

There are five major modes of life adopted by animals generally:
1. **Terrestrial**—living mainly on the land.
2. **Aquatic**—living largely in water.
3. **Arboreal**—living largely in trees.
4. **Aerial**—living largely in the air.
5. **Fossorial**—living largely underground.

Mammals have become adapted to all these major modes of life. Thus, although the earliest mammals which existed some 70 million years ago were mainly terrestrial and the majority of modern species still are, in the course of evolution some have taken to an aquatic life (whales, seals and otters), others have become arboreal (monkeys, squirrels and lemurs), some have become aerial (bats), and others fossorial (moles). When a single group of animals spreads to many habitats and becomes adapted to several modes of life, the process is known as **adaptive radiation**. Mammals show this type of evolution remarkably well.

Some examples of adaptation

It is not surprising that life in these very different habitats has resulted in major adaptations of structure. This is particularly true of the limbs, as each mode of life requires a modified means of locomotion. It is remarkable that the characteristic pentadactyl limb has become adapted so that it may be used efficiently, for example, by deer for fast running on land, by whales as flippers for swimming, by squirrels for jumping and gripping boughs, by bats for flying, and by moles for shovelling earth and burrowing in the ground. By studying the skeletons of their limbs (Fig. 11:3) you will see how each has been modified from the primitive pentadactyl arrangement. The adaptations that have taken place have not only affected the limbs, but other parts too. For example, different senses become more important than others in different habitats,

Fig. 11:2 Adaptive radiation in mammals. a) Aerial: bat. b) Arboreal: grey squirrel. c) Terrestrial: red deer. d) Fossorial: mole. e) Aquatic: white whale.

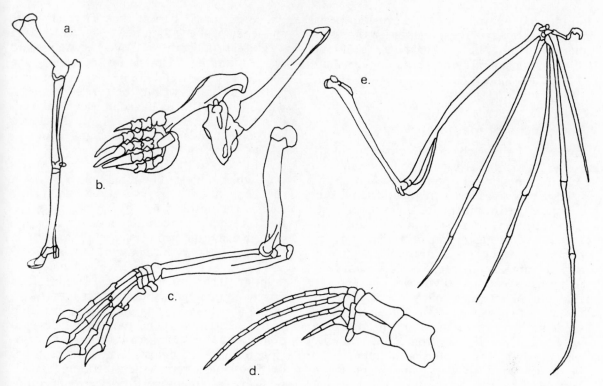

Fig. 11:3 Diagram showing how the fore limbs of certain mammals have become adapted for different purposes. a) Deer: running. b) Mole: digging. c) Squirrel: jumping and climbing. d) Whale: swimming. e) Bat: flying.

TABLE SUMMARISING SOME OF THE ADAPTATIONS OF MAMMALS TO DIFFERENT MODES OF LIFE

	Terrestrial (Deer)	Aquatic (Whale)	Arboreal (Squirrel)	Aerial (Bat)	Fossorial (Mole)
Locomotion	Long legs, runs on its toe-nails (hooves); good for gripping.	Forelimbs as flippers. Very strong tail with fin.	Long hind limbs for jumping. Long fingers and claws for climbing and grasping. Long tail as balancer.	Forelimbs as wings.	Forelimbs very strong and broad for digging.
Senses	Good eyes, ears and nose to detect predators in time to escape.	Uses type of sonar for communication and navigation.	Good eyesight for judging distances when leaping.	Sonar for navigation in dark.	Sensitive vibrissae to detect vibrations in ground. Good nose for detecting prey.
Feeding	Special type of tongue, mouth and teeth for grazing and browsing.	Enormous mouth. Baleen for filtering plankton.	Long fingers for holding food while gnawing.	Sonar for catching food; tail membrane acts as net for catching insects.	Makes system of tunnels which act as traps for its prey.
Reproduction	Young can run very soon after birth.	Young can suckle in water.	Young can climb even before their eyes open.	Young can be carried by mother when flying.	Young protected in nest below ground.

and feeding and reproduction provide special problems. Some of these adaptations are summarised in the table on p. 113 and you could add to them considerably by reading more about the species concerned. We will study certain aspects in more detail.

The senses of mammals

In birds, sight is the most important of the senses although hearing is also very acute; the sense of smell is very poorly developed. This is largely because birds are mainly active during the day (**diurnal**). The majority of mammals, however, are most active at night (**nocturnal**), or during the transition period around dusk and dawn (**crepuscular**). Consequently, many mammals rely more on scent and hearing. The exceptions, as you would expect, are the diurnal ones. It follows from this difference between birds and mammals that when studying birds in the field you need to keep very still or watch them from a hide, but with many mammals it is often more important to be down wind so that your scent is not detected.

Bats have a special problem due to their specialised mode of life. They are nocturnal or crepuscular, and many of them feed on insects such as moths, beetles and mosquitoes. They have to catch their prey while on the wing and at the same time avoid damaging themselves against boughs and other obstructions in their way. They do this by a remarkable echo-locating system known as **sonar**. This involves the giving out of rapid pulses of ultrasonic sounds which rebound off any object in the bat's flight path and are picked up by the extremely sensitive ears of the bat. This method of detecting objects is so effective that if bats are released in a room where extremely fine wires have been strung across no more than a wingspan apart, they will avoid striking them even in complete darkness.

In some bats the sound pulses are emitted through the mouth at a rate of about 10 per second when cruising, but if an insect such as a moth is detected the rate is increased to more than 100 per second which enables the bat to investigate it more carefully and pin-point its position. In the leaf-nosed bats the sounds are emitted through the nostrils which serve to concentrate them in a narrow beam like a bullet in a rifle barrel, thus increasing the efficiency of the technique.

The ears of bats are relatively very large and are admirably adapted for receiving the echoed sounds. In addition, they have a projecting structure within the ear, the **tragus**, which helps to focus the sounds on to the ear drum. The inner ear is also highly developed to receive sounds of extremely high frequency, and by some means not yet understood, bats are able to hear the echoes of their own sounds even when subjected to much louder noises at the same time. They can also avoid hitting against each other when flying in great numbers together in the confines of a cave, although each is using its own sonar at the same time. Can they, perhaps, recognise their own voices?

The aquatic mammals, such as whales and dolphins, have similar problems of finding food and avoiding objects, but in water rather than in the air. This is particularly true in turbid water or at night. The toothed whales and dolphins use a sonar device similar in principle to that used by bats. As with bats, it has been possible to record these ultrasonic sounds by using extremely sensitive electronic equipment. When they are played back at a reduced speed, they are audible to the human ear as a series of clicks.

Whales also have the problem of communicating with each other over great distances of apparently featureless ocean. They do this by uttering a wide range of whistles, groans, grunts and trills which are conducted through the water over very considerable distances and are apparently meaningful to other whales. It is even thought that individual whales may be recognised by others by their signature tunes!

The characteristics of mammals

Let us now consider some of the features which distinguish mammals from other vertebrates, and which have been largely responsible for their success.

Hair
All mammals have some hair although some species such as whales only show a few bristles as adults. In fur-bearing types hairs are of two kinds, the long **guard hairs** and the thick felt-like mass of finer hairs forming the **under-fur**. The hair traps air next to the skin,

Fig. 11:4 Long-eared bat.

insulating the body from too much loss of heat. In otters and polar bears, for example, the underfur is so thick that when they are swimming the water cannot penetrate and the skin remains dry.

Hair has other functions beside preventing loss of heat. Some hairs are modified as special sense organs of touch, e.g. the **vibrissae**, or whiskers. In otters these vibrissae are particularly sensitive; they can detect the vibrations in the water made by a fish swimming, enabling the otter to locate the position of the fish and catch it even in dark or turbid waters.

In hedgehogs and porcupines the hairs have become modified as spines or quills and are defensive in function. If you examine a hedgehog you will see how the spines of the back and sides merge into the normal hairs of the underside. The horns of the rhinoceros are also made of hair, not bone as you would expect. They consist of a mass of hair impacted together to form a very strong and formidable weapon.

The colour of hair in different mammals is often protective, enabling the animal to blend with its surroundings. In the majority the fur is darker above and lighter below, a condition known as **counter-shading**. As dark colours reflect less light than pale, and most light comes from above, counter-shading tends to equalise the amount reflected from all parts and so gives an appearance of flatness which makes the animal far less conspicuous. Thus a rabbit is difficult to see when it is feeding in a field, but a dead one, lying on its side, is immediately visible.

In other mammals the outline is broken up by stripes or spots as in tigers and leopards, or by having a single conspicuous dark line along the side as in many antelopes. This type of camouflage is known as **disruptive colouration**. It is very effective because it takes the eye away from the outline of the animal which gives it its familiar look.

Other mammals which live in places where there is one main prevailing colour often have fur of the same colour. Desert mammals are usually some shade of yellowish brown and those in snowy regions are white. This gives an advantage both to a predator when hunting and to the prey when being hunted. Polar bears and the young seals on which they feed are both white, and the same applies to the Arctic foxes which hunt the Arctic hares.

Some mammals change the colour of their coat according to season. This is true of the northern race of the stoat, which in winter has a white coat and is known as **ermine**, but in summer has one that is reddish-brown. Arctic foxes and Arctic hares undergo a similar change, but the new coat is acquired differently. In the fox the dark tips of the hair wear off as winter approaches leaving only the white basal parts which continue to grow; it then moults this coat in the spring, replacing it with the summer one which is blue-grey. Arctic hares, on the other hand, effect the changes by having several moults.

One of the strangest examples of hair colouring is in some of the South American sloths which live high up in the trees; they have green hairs. This is because microscopic green algae are living in the groove which runs down each hair. The general effect is to make them most inconspicuous.

Viviparity

All but the platypus and echidna (which lay eggs) bear young, and are thus said to be **viviparous**. This differs from the viviparity of other vertebrates we have mentioned such as sharks and adders, because in mammals the young are fed before birth through an organ called the **placenta** from which food is extracted from the mother's blood. When fish and reptiles bear young, the developing animals feed only from the yolk inside the egg,

Fig. 11:5 Different kinds of hair in mammals.
a) Otter: vibrissae for detecting vibrations in the water.
b) Bear: fur for warmth. c) Hedgehog: spines for defence.
d) Rhinoceros: the horn is made of compacted hair.

Fig. 11:6 Representatives of some of the main orders of mammals. a) Elephant (Proboscidea). b) Kangaroo (Marsupialia). c) Chimpanzee (Primates). d) Hare (Lagomorpha). e) Shrew (Insectivora). f) Wood mouse (Rodentia). g) Fox (Carnivora). h) Giraffe (Artiodactyla). i) Zebra (Perissodactyla). j) Leopard (Carnivora).

and the mother merely keeps the eggs inside her body until they hatch; they are then born. This method is distinguished by the term **ovoviviparity**. Marsupial mammals, e.g. the kangaroo, are somewhat intermediate between these two types as the young are born in a very immature state. They are then suckled inside a pouch or **marsupium** which also protects them.

Secretion of milk

All female mammals secrete milk and suckle their young after birth. In all but the platypus and echidna the milk is secreted from **mammary glands**, also called breasts or udders in certain species. These vary both in number and position. In man and elephant there is only one pair which is thoracic, in others such as cats, cows and pigs there are two or more pairs which are abdominal. Those mammals which have large litters tend to have more. Suckling enables the mother to provide food for its young until it is **weaned**, i.e. until it can feed on solid food.

A glandular skin

Hairs, like feathers, need to be kept supple and waterproof. Mammals have minute oil glands in the skin which lubricate the base of each hair. They also possess sweat glands.

Teeth

Mammalian teeth, unlike those of the lower vertebrates, develop in sockets in the jaw. Usually they are of four kinds: incisors in front for cutting or gnawing, behind them the canines for killing the prey or tearing the food, and further back the premolars and molars for

Fig. 11:7 A sow suckling.

chewing or grinding. The dentition of different species varies very much according to the type of food eaten, a feature which is used in the classification of the different mammalian orders. We shall study teeth in detail in Chapter 17.

Other mammalian features

There are other characteristics which mammals share with other vertebrates, but which are developed to a greater degree. One of these is the brain which reaches its highest development in the whales and dolphins, and in apes and man. This has enabled mammals to learn more about their surroundings, develop better memories, and behave with greater intelligence than other vertebrates.

Parental care has also been developed to a far greater degree. Not only do they feed and protect the young in the uterus before birth and suckle them after birth, but they also protect them during the growing-up period. Consider the essential role of parental care in our own lives!

Projects you could carry out on mammals

1. Study a species of mammal for yourself, e.g. dog, cat or mouse, or wild ones such as rabbit, fox or badger. Consider first its mode of life, where it lives, what it feeds on and its general habits. Then try to discover in what ways it is adapted to live in such a habitat, feed, move and breed in the way it does. Whatever species you choose, whether it is a cat living in a city suburb, or a badger in a wood, you will find it is well adapted both in structure and behaviour to its mode of life.
2. Collect as many kinds of animal hairs as you can and try to identify their owners by the colour, length, curliness, etc. Hair often gets caught in the barbed wire round fields. Where there is a barbed wire fence bordering a wood, look especially at the lowest strand where an animal path leaves the wood. Brambles and briars also collect hairs; look carefully at those bordering animal paths.
3. Observe the behaviour of some of the small mammals such as mice and voles. Find out which species occur in your garden or in a hedgerow by attracting them with food. Just

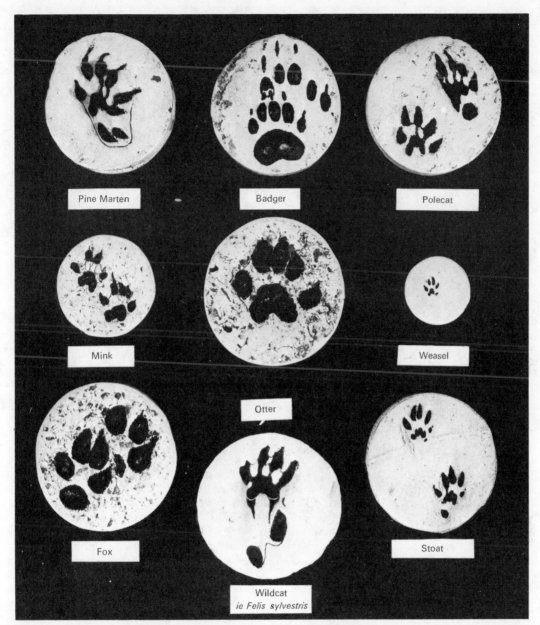

Fig. 11:8 Plaster casts of the footprints of some mammalian carnivores.

before dark put out small piles of oat grains or vegetable scraps in sheltered places where birds will not find them. Put more down each night and examine each morning. You will probably find that the food disappears quite regularly after a few nights. Discover what is eating the food by watching at night using a torch with a red filter to help you. (Many mammals are insensitive to red light.) Make detailed notes of what you see.

4. Look out for mammal tracks in snow or in muddy places. Look especially in the mud of woodland paths, by rivers, and near the edges of ponds and lakes. Identify the tracks with the help of a reference book. In dry weather you can make your own mud at strategic places and peg down some clean paper near it for the animals to tread on after going through the mud.

Take any opportunity when snow is on the

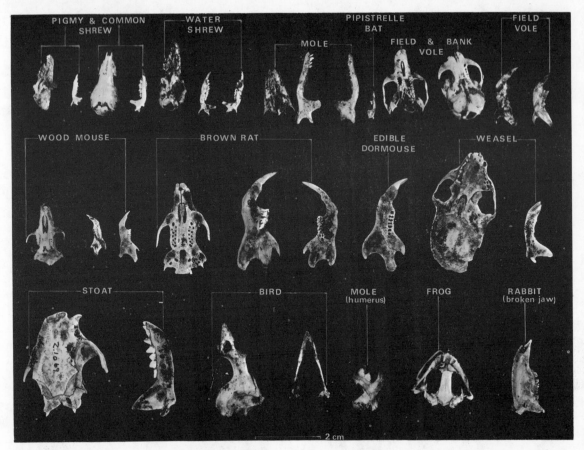

Fig. 11:9 Some skeletal remains from owl pellets.

ground to follow the tracks of mammals. In this way you can discover much about the habits of foxes, rabbits, badgers and many others.

5. Make a collection of plaster casts of the best footprints you find. To do this take a plastic strip and make it into a ring a little bigger than the print, fastening the ends together with a paper clip. Press the ring gently into the ground round the print and pour in a runny mixture of plaster of paris and water, made up on the spot. You can remove it carefully after about ten minutes when the plaster should have set, but it should not be cleaned up until later. To do this, place the cast under running water to remove the mud. Be sure to label it on the back with date, locality and, of course, the scientific name of the mammal.

6. Find out what small mammals exist in your neighbourhood by examining owl pellets. Owls are expert at catching them, but having digested the soft parts they eject the bones and fur in a pellet through the mouth. These may be found under places where owls roost. Soak the pellets overnight, and then pick out the skulls and bones with forceps. Identify their owners with the help of a reference book. In this way you may find the skulls of various species of shrews, mice, voles and even an occasional mole or bat. The parts of the skeletons can be cleaned in water and bleached by soaking in hydrogen peroxide solution and then mounted on cards.

7. Another method of finding out about the distribution of small mammals is to examine the contents of old bottles which have been thrown away near lay-bys and similar places by careless people. Small mammals often enter these bottles, are unable to get out and so die. Wash out the contents of the bottle into a dish and look for any skeletons. The bones can be cleaned, bleached and mounted as in 6. Put the bottles in a litter bin!

12

Diffusion and osmosis

In previous chapters we have been largely concerned with the study of whole organisms: their great diversity, their life histories, modes of life, and the ways in which they are adapted to living in various environments. We have also seen how all these organisms, whether plant or animal, large or small, carry out certain functions which are common to all living things; they all feed, respire, excrete, respond to stimuli, grow and reproduce. We shall now study more deeply these functional aspects of biology—how the body of an organism really works. For this we shall concentrate mainly on man and flowering plants: man, because he is the most important and interesting member of the animal kingdom to us; flowering plants, because we are utterly dependent upon them for life, and need to understand the contrasting ways in which they carry out their life functions.

As an introduction to this study, it is necessary first of all to understand two physical processes which are of vital importance to all living organisms. These processes, **diffusion** and **osmosis**, are concerned with the manner in which water and other substances move within the tissues of organisms.

Diffusion

Consider a solution of sugar in water. It consists of water molecules of extremely small size and sugar molecules of rather larger size. All these molecules are moving rapidly and in a random manner, hence they are evenly distributed within the liquid.

When there is a greater concentration of sugar in one part of the solution, as would happen when a lump of sugar slowly dissolves in it, more molecules tend to move away from the region of high concentration than return to it, until eventually they become uniformly dispersed once more. This movement of molecules (or the ions into which they dissociate when in solution) from regions of high concentration to low, until uniform dispersal results, is called **diffusion**. The difference in concentration between the two regions is known as the **diffusion gradient**. Diffusion occurs in both gases and liquids.

You can observe diffusion taking place by using a substance such as potassium permanganate which forms a coloured solution in water. The density of the colour is an indication of its concentration.

Place a beaker, three-quarters full of water, in a place where it will be undisturbed for several days. Introduce a little strong potassium permanganate solution so that it forms a concentrated layer at the bottom of the beaker. You can prevent too much mixing by pouring it very slowly down a glass funnel held firmly in position by a clamp; the end of the funnel should touch the bottom of the beaker. Note any change in colour of the liquid. Does it become uniform? If so, how quickly does this happen?

Diffusion is a very important process in living organisms as substances such as oxygen, carbon dioxide, salts and sugar will tend to move from regions of high concentration to low. For example, if cells are constantly using up a gas such as oxygen, the concentration of oxygen there will be lowered, hence more oxygen will diffuse from places of higher concentration to take its place. In this way a constant supply to the cells can be kept up. We shall come upon many other important examples of diffusion in other chapters.

Osmosis

An important variation of diffusion may occur when there is some kind of barrier separating the solution of high concentration from the low. What happens depends on the kind of barrier. If it is made of something such as

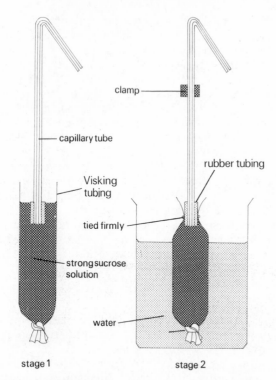

clamp

capillary tube

Visking
tubing

rubber tubing

tied firmly

strong sucrose
solution

water

stage 1

stage 2

Fig. 12:1 Osmometer.

below the level of the solution and tie firmly to
make a watertight joint with the capillary
tube. Wash the outside of the bag to remove
any spilt sugar solution and suspend it in a
beaker of water. Mark the level of the sugar
solution in the capillary and observe any
changes over a period of half an hour.

It is possible to explain any rise in level by
saying that water has passed from the beaker
through the Visking tubing into the sucrose
solution thus increasing its volume. If you
think of this in terms of movements of molecules (Fig. 12:2) you can say that water has
passed from a region of 100% water through
a semi-permeable membrane into a sugar
solution where the water is less than 100%;
thus the water molecules have obeyed the
diffusion law by going from high concentration to low, but the sugar molecules have been
prevented from doing so because they could
not pass through the pores of the membrane.
This process is called **osmosis** and it also
takes place when a semi-permeable membrane
separates a strong solution of sucrose from a
weaker solution—it does not have to be pure

Fig. 12:2 Diagram to illustrate the action of a semi-
permeable membrane. (Small circles = water molecules,
larger circles = sugar molecules). Note that more water
molecules pass from right to left than in the reverse
direction.

glass, no molecules can move through it,
hence it is said to be **impermeable**. If it is
made of a porous material with pores sufficiently large to let the largest molecules
through, diffusion will occur through it as if
it did not exist; it is then said to be fully
permeable. The cellulose walls of plant cells
are like this. However, if it contains pores
which are of molecular size, it could be that
small molecules such as water will be able to
pass through, while larger ones such as sugar
will not. Such a substance is said to be **semi-
permeable**. In other words, by being selective in its action, the membrane modifies the
diffusion process.

Visking tubing is an artificial semi-permeable membrane which is convenient to use for
studying the process.

Set up the apparatus shown in Fig. 12:1
using a piece of Visking tubing about 10 cm
long. First wet it and then tie one end in a knot.
Half fill it with strong sucrose solution and
insert the bent capillary tube. Now remove the
air by squeezing the Visking tube tightly just

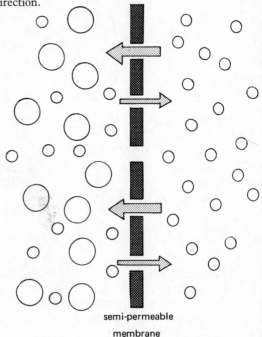

semi-permeable

membrane

water. More water molecules pass from the weaker solution to the stronger than in the opposite direction.

From your experiment you will have noticed that the intake of water causes a rise in the capillary tube against gravity; it causes a pressure sufficient to force it out at the top. This is called the **osmotic pressure** of the solution. Its value is equal to the pressure that would have to be exerted in the opposite direction on the column in the capillary tube to prevent it from rising. The stronger the solution, the greater is the osmotic pressure.

Osmosis in living cells

We can now find out whether osmosis occurs in living material and if so whether the semi-permeable membrane is living or non-living. For this we will compare the action of living and dead potato tissue.

Peel a large potato, cut it into two halves and remove the central tissue to form two cups (Fig. 12:3). Boil one half in water for about 3 minutes to kill the cells and then suspend each half in a beaker. Add enough strong sucrose solution to each half potato to half-fill the cup, and put enough water in the beaker to come up to the same level. Note the levels in the two potatoes after 24 hours. Have the living cells acted as an osmotic system, causing water to be drawn into the strong sucrose solution? Does it make any difference whether the cells are living or dead? What do you think would have happened if you had put water in the potatoes and sugar solution in the beakers?

It is believed that the semi-permeable membrane which causes living material to be an osmotic system is the surface membrane of

Fig. 12:3 Experiment to find out whether living material acts as an osmotic system.

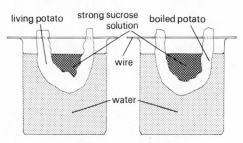

the living cytoplasm—the **plasma membrane**. This membrane is present wherever cytoplasm and water meet and is thought to be only a few molecules thick. It is very selective as to which molecules pass through, but it does *not* act quite like the rigid sieve we considered in order to explain how a semi-permeable membrane worked. In practice some large molecules do pass through. It is easy to show this with the potato experiment by testing for sugar (p. 161) in the water in the beaker containing the living material after a few days. What probably happens is that the larger molecules pass through much more slowly than the water molecules, so the osmotic effect is the same. Other theories to explain osmosis have been put forward, but these will not be considered here.

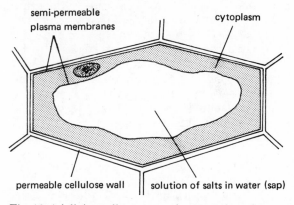

Fig. 12:4 A living cell—an osmotic system (membranes shown diagrammatically).

If you think back to your observations on living plant cells (p. 5), you will see that each cell is an osmotic system (Fig. 12:4). The cytoplasm surrounds a vacuole containing water in which substances such as salts and sugars are dissolved, and both internal and external surfaces of the cytoplasm will act as semi-permeable membranes. What would happen if cells such as these were surrounded by solutions of various strengths? It is difficult to observe the effect in some cells as the cytoplasm and cell sap are colourless, but you could use cells where the sap is coloured, as it is in the epidermal cells of plant organs which appear red.

Strip a small piece of epidermis from the reddish area of a rhubarb petiole and mount it on a slide in a drop of water with the outside uppermost. Examine under the microscope and note the red content of some of the cells. Now remove the water with blotting paper and add a few drops of strong salt or sugar solution and note what happens to the red content of the cells. If you observe a change, remove the strong solution once more and replace with water. Is there any further change? Do your observations confirm that living cells act as osmotic systems?

The changes you should have seen may be explained more fully by saying that when living cells are placed in a **hypertonic** solution (one stronger than that of the cell sap), water will be drawn out from their vacuoles and the cytoplasm, which is somewhat elastic, will come away from the cell wall. At the same time the bathing solution will pass through the cell wall, because it is permeable, and fill the gaps. In this condition the cell is said to be **plasmolysed**. When the cell is then surrounded by water or a **hypotonic** solution (one weaker than the cell sap), water will be drawn into the vacuole which will expand until the cytoplasm is forced back against the cell wall. But the cell sap will still be stronger than the fluid outside, so more water will be taken in and a pressure will be exerted outwards against the cell wall causing it to stretch. The cell is now said to be **turgid**. This will continue until the increasing resistance of the cell wall to this stretching equals the osmotic effect of the water being drawn in.

Bearing these principles in mind, carry out this experiment:

Cut from the rhubarb petiole used in the last experiment two strips of tissue (with any epidermis removed) about 8 cm long and roughly 0·5 cm square in section. Use a ruler to guide the scalpel when making the cuts. Trim the ends so that they are of equal length and note the exact measurement. Now place one strip in strong sugar (or salt) solution and the other in water and leave for half an hour. Now measure each strip again carefully. Is there any difference in the lengths? Feel the two strips and compare their rigidity. Is there any difference between them? Can you explain any difference in terms of osmotic action? (This experiment can be done equally well using cylinders cut from a large potato by means of a cork borer.)

You will now realise from your experiments that it is osmosis which causes water to be drawn into plant cells causing them to swell and press against each other so bringing about the rigidity of stems and leaves (p. 52). It is also by osmosis that water is drawn into the root hairs of plants from the soil.

What happens to animal cells under similar circumstances, as they have no tough cell walls to prevent them from expanding? You could find out by subjecting cells such as red blood corpuscles to solutions of different strengths. It is easy to obtain the material as each drop of blood contains about 5 million red corpuscles. These corpuscles normally float in a fluid plasma which contains various salts and sugars in a concentration which gives it the same osmotic strength as would be obtained by dissolving 0·85 g of salt in 100 cm^3

Fig. 12:5 a) Cell in hypertonic solution—plasmolysed. b) Cell in hypotonic solution—fully turgid.

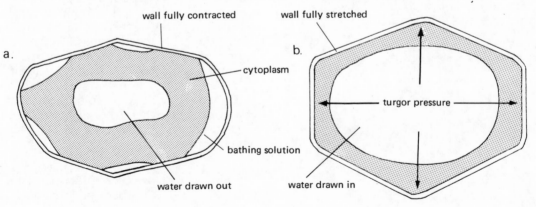

124

of water. Solutions of equal osmotic strength are said to be **isotonic**. So for this experiment we shall need:
1. A 0·85% solution of sodium chloride in water which is isotonic.
2. A 2% salt solution which is therefore hypertonic.
3. Distilled water which is hypotonic.

Place 1 cm^3 of each of the above solutions into 3 small specimen tubes, appropriately labelled, and add a drop of blood to each. To obtain the blood sterilise the top of a finger with surgical spirit. Draw more blood into your finger by swinging your arm round, holding it downwards and wrapping a handkerchief tightly round the base of the finger. Now make a quick jab with a sterilised lancet and draw up a few drops of blood into a fine pipette which has also been sterilised. Place one drop into each of the specimen tubes and shake gently. After a few minutes remove a drop from the isotonic solution with a clean glass rod and examine under the high power of the microscope. Note the shape of the corpuscles and remember this is your control. Now do the same for the hypertonic and hypotonic solutions. Observe what happens to the corpuscles in each case. Is there any sign of the corpuscles having burst? Does any difference in the appearance of the fluids in the three tubes confirm your observations and deductions?

From this experiment you should appreciate the importance of not allowing animal cells to be bathed in fluids which are much stronger or much weaker osmotically than that of their cytoplasm.

Can you now explain the action of a contractile vacuole in an amoeba which lives in pond water, and why amoebae living in sea water do not have one?

We can now summarise the important aspects of diffusion and osmosis. When diffusion occurs molecules in gases or liquids pass from regions of high concentration to low until there is an even dispersal of molecules. Diffusion is an important process whereby substances such as salts and sugar when dissolved in water, and gases such as oxygen and carbon dioxide travel within the tissues of plants and animals.

Osmosis in living systems only concerns the passage of water. The water passes from dilute solutions to more concentrated ones because of the selective nature of the semi-permeable membrane. The membrane that acts in this way in living things is the surface membrane of the cytoplasm.

13

Obtaining energy

All animals and plants need energy to carry out the vital processes going on in their bodies. We need energy to move, to talk, to think, to grow; we need it when asleep or awake, from conception to death. Every living cell in our bodies needs energy. How do organisms obtain this energy? The short answer is through **respiration**. This essential life process may be defined as the liberation of energy from food. To find out how this happens let us first make a comparison between respiration and a process we all know something about— **combustion** (burning). When a fuel such as coal, wood or gas burns, energy in the form of heat and light is liberated. This process only takes place when oxygen is present, and as a result the fuel is used up and two of the products which pass into the air are carbon dioxide and water vapour. The equation for combustion can be written:

$$\text{fuel} + \underset{\text{oxygen}}{O_2} \rightarrow \underset{\substack{\text{carbon} \\ \text{dioxide}}}{CO_2} + \underset{\text{water}}{H_2O} + \text{energy}.$$

You will know that certain foods we eat are called energy foods; sugar is a good example. If sugar is burnt in a crucible in the presence of a lot of oxygen it behaves like a fuel: it bursts into a bright flame, gives out a lot of heat, and carbon dioxide and water are formed. Is respiration something like this?

To find out we shall have to consider five experiments on living material, each concerned with one aspect of the above equation. We must show that:

1. energy is released,
2. carbon dioxide is given out,
3. water is formed,
4. food is used up in the process,
5. oxygen is used up.

1. When organisms respire, is energy released in a form we can detect?

The easiest form of energy to detect is heat, so see if you can measure any heat given off by such things as seeds when they are germinating.

Prepare some germinating oats by soaking a handful of dry oats in water for 24 hours. Kill a similar sample by soaking in 10% formalin. Drain excess fluid from both. Rinse the living seeds in a dilute solution of some mild disinfectant such as TCP which will kill off any bacteria on the surface of the seeds, but leave the seeds unharmed. Why should this precaution be taken? Now set up the apparatus as in Fig. 13:1. The dead seeds act as a com-

Fig. 13:1 Apparatus to determine whether heat is released from living material.

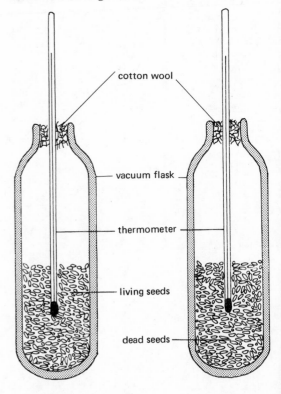

cotton wool

vacuum flask

thermometer

living seeds

dead seeds

parison or **control**. A control is usually essential in biological experiments to ensure that we interpret the results correctly. Both experiment and control are similar and they must be kept under the same conditions except for the one factor which is being investigated.

Think out:
1. Why it is better to have a *lot* of oats in each.
2. Why cotton wool is used instead of a cork.
3. Why vacuum flasks are used instead of glass flasks.

Compare the temperatures each day for several days. What do you conclude from your results? If you put your hand into a heap of lawn mowings a few hours after they have been cut, the heap feels warm. Can you see any connection between this and the result of your experiment?

2. Is carbon dioxide given out when an organism respires?

There are two good tests for carbon dioxide. One is to pass it into limewater which then turns milky, the other is to pass it through bicarbonate indicator which gradually changes colour as the concentration of carbon dioxide increases. (This is because the solution becomes more acid.) The colour change is from orange-red to yellow. First, use one of these tests to discover if you give off carbon dioxide yourself when you breathe out.

Set up the apparatus shown in Fig. 13:2 and breathe *slowly* in and out. As you breathe in, the air passes through the left-hand tube and when you breathe out it goes through the right. (Notice the position of the glass tubing which controls the direction of airflow.) Why is there an equal volume of indicator in each tube?

Note any change of colour and conclude whether or not the exhaled air contains more carbon dioxide than the inhaled air.

Do other organisms breathe out carbon dioxide? To find out, use the apparatus shown in Fig. 13:3. Various members of the class could use different material such as woodlice, beetles, locust nymphs, germinating peas, chopped-up potato, etc.

Rinse out two boiling tubes with distilled water followed by a little bicarbonate indicator. Label them A and B.

Add 2 cm^3 of indictor to A and cork it at once (avoid breathing into it). This acts as a control.

Now add 2 cm^3 of indicator to B and put in the cork with the rod and perforated zinc disc attached; introduce the living material before fixing the cork tightly.

Fig. 13:2 Method of testing whether we give out carbon dioxide when we breathe.

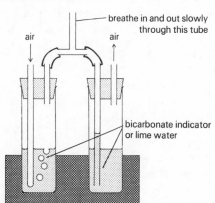

Fig. 13:3 Apparatus to test whether living things give out carbon dioxide.

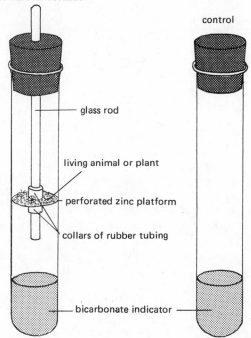

Shake the solution at intervals, but do it gently so as not to splash the living material. Note if the colour changes compared with A.

Do all the specimens give out carbon dioxide?

This apparatus can also be used to compare the *rate* at which carbon dioxide is given out when the temperature is altered. You could work out the details of how this could be done remembering that when you alter the temperature everything else must be kept exactly the same if the comparison is to be accurate. You will need a third tube C containing bicarbonate indicator solution which you have already blown through to produce a standard yellow colour to act as an end-point.

3. Is water given out when an organism respires?

It is easy enough to test your own breath for water vapour with thoroughly dried cobalt chloride paper (which turns from blue to pink), and you know how this water vapour condenses in cold weather so that your breath becomes visible, but it is difficult to prove that this water actually results from the breakdown of the food, as all living things contain water anyway. However, scientists have proved this by using radioactive isotopes. These are atoms which can be incorporated into the molecule of a compound such as sugar which is then said to be 'labelled'. The atom can now be detected wherever it goes, because although it behaves like a normal atom, it also gives off bursts of radiation. These can be detected by electronic Geiger-Muller tubes which are connected to a machine which counts any bursts of radiation detected by the tubes. In this way it has been shown that if glucose sugar containing radioactive *hydrogen* atoms is fed to animals, the water in their breath becomes radioactive. By the same technique using radioactive *carbon* atoms in the glucose it has been shown that the carbon dioxide in exhaled breath is radioactive. Hence both the hydrogen in the water and the carbon in the carbon dioxide have come from the sugar.

4. To show that food is used up when an organism respires

The easiest way to show something is used up is to weigh it before and afterwards. But with a living organism this is difficult as its weight is altering all the time because a) it is taking in or making more food, b) its water content is altering. To overcome a) you can deprive it of any source of food; this is easy to do with seeds which are germinating because the food is already stored in the seeds, and to make them germinate you need only give them water. Also, by keeping them in the dark you prevent them from making any more food. To overcome b) the *theoretical* way is to eliminate all water from the seeds at the start and record their dry weight; then give them water to allow them to germinate and after several weeks of growth remove all the water again and record the dry weight once more. But to eliminate the water, you have to put them in an oven at 100°C until the weight is constant, and if you do this you kill the seeds! That is the problem! You can only overcome this by taking two lots of seeds which are equal in number and weight. You assume that the water content of both lots is the same and reduce one lot to dry weight. Proceed as follows:

Take 10 broad beans, put 5 on one pan of a balance and 5 on the other and change them around until their weights are equal.

Put one set in an oven and reduce to dry weight (you assume this equals the dry weight of the other). To obtain an accurate dry weight reading, cool the seeds in a desiccator before weighing them otherwise they may absorb water from the air. Return the seeds to the oven and repeat the procedure until you obtain a constant reading. Make a note of this weight.

Plant the second set in a bowl of wet sawdust and leave in the dark for at least two weeks so that they will grow and respire.

You now have to dig up the plants without breaking the roots, and remove all the sawdust which adheres to them. Do this by first soaking the bowl in water and then washing each plant gently under the tap.

Put the plants in an oven and reduce to dry weight.

Compare this weight with the dry weight of the first set of seeds. If it is less, what do you conclude?

The experiment is based on the assumption

that the two lots of seeds were alike. Would this assumption have been more justified if you had taken many more seeds? (Compare your results with those of the whole class.)

5. To show that oxygen is used up when an organism respires

Set up the apparatus as in Fig. 13:4. The experiment is based on the principle that if oxygen is used up there should be a reduction in the volume of air in the boiling tube. But we have already shown that living organisms give off carbon dioxide, so this additional quantity of gas could mask the reduction of oxygen. It is therefore necessary to absorb the carbon dioxide as soon as it is formed by putting in a small tube of soda lime.

The second boiling tube containing dead seeds is needed to act as a control, because respiration is a living process. By having tubes of the same capacity any change in volume due to temperature variation would affect each equally.

Fig. 13:4 Experiment to find out whether oxygen is absorbed when an organism respires.

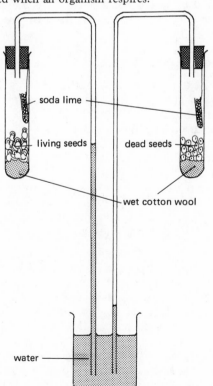

soda lime

living seeds   dead seeds

wet cotton wool

water

See that the levels in the two glass tubes are the same at the start. Leave for 3 or 4 days and note any change in level in the two tubes. If there is a change you still have to prove that the gas absorbed is in fact oxygen. To do this remove the cork from the first boiling tube and slowly lower a lighted taper into it. It will go out at once if little or no oxygen is present. Test the control in the same way.

From these experiments we can now conclude that when an organism respires the process appears to be comparable with that of burning sugar in oxygen.

$$\text{sugar} + \text{oxygen} \rightarrow \frac{\text{carbon}}{\text{dioxide}} + \text{water} + \text{energy}.$$

As organisms release energy mainly from sugars and fats, and glucose sugar is one of the most important, we can write the respiratory equation more fully:

$$\begin{array}{cccc} & & \text{carbon} & \\ \text{glucose} & \text{oxygen} & \text{dioxide} & \text{water} \\ C_6H_{12}O_6 + & 6O_2 & \rightarrow 6CO_2 & + 6H_2O + \text{energy} \\ \text{(containing} & & & \text{(released)} \\ \text{trapped} & & & \\ \text{energy)} & & & \end{array}$$

But there must be some difference between combustion of sugar by burning and respiration in living cells, because:
1. In the laboratory, sugar must be brought to a high temperature before it burns and liberates its energy. If this happened in our cells we would become charred!
2. Once the sugar starts burning we cannot stop it easily, but living cells seem to be able to control the process.
3. Although cells contain a lot of water, respiration goes on within them; outside the body water normally stops combustion from taking place.

Therefore we must conclude a) that respiration is a very different process from combustion, although the same substances are used and the end products are similar; b) that a living cell is somehow able to control the release of energy; c) that the reaction takes place at normal temperatures and in a watery medium.

The full explanation of how this happens is

very complex, but some of the important points are these:

1. The sugar is broken down into carbon dioxide and water by a complex series of reactions involving a number of intermediate compounds.
2. Some of these reactions release energy, but always in small quantities.
3. The reactions occur at normal temperatures because each is controlled by an **enzyme**.
4. The energy is liberated within the cell when it is needed for some vital process.

Enzymes

These are complex chemicals which are made by the cytoplasm of the living cells; they control the chemical reactions. Enzymes act as **catalysts,** i.e. they speed up a reaction, which otherwise would be extremely slow at normal temperatures, and at the end of the reaction they remain unchanged, so they can be used again and again. Each reaction is catalysed by a different enzyme.

The way an enzyme works is rather like the way a key opens a lock. The key has to be of the right pattern to perform the action, but once it has acted it can be taken out and used on any other similar lock. An enzyme, in the same way, has a particular molecular shape which only reacts with a substance whose molecule it fits. Thus the series of reactions bringing about respiration are controlled by a series of different enzymes.

Forms of energy

Some energy is released in the form of **heat** as we have already seen; occasionally it is in the form of **light**. Glow-worm beetles, for example, give out light which enables the males to find the females prior to mating. Many deep-sea fish have luminous organs, and certain bacteria and fungi may be seen in the dark because of the light energy they emit. But the energy required by living cells, for such processes as the contraction of muscle, needs to be in a different form. It must be readily released when required. How does this happen?

When energy is released from the food during respiration, some of it is transferred to a special substance found in all cells called **ATP** (adenosine *tri*phosphate). One of three phosphate groups which form part of the ATP

molecule is easily removed to form ADP (adenosine *di*phosphate), thus releasing energy from the molecule. Energy from the breakdown of food is used to recombine the ADP and phosphate to form ATP once more.

1. ATP $\longrightarrow$ ADP + phosphate + energy.
2. Energy + phosphate + ADP $\rightarrow$ ATP
   (from food
   breakdown)

By using a specially prepared ATP solution it is possible to demonstrate this action by putting a drop on a length of fresh muscle fibre from lean meat—it will immediately contract. Thus the **chemical** energy stored in the ATP has been turned into **mechanical** energy, i.e. for doing work. When ATP releases its energy not all of it is used for contracting the muscle. Some is lost in the form of heat; that is why, when you use your muscles a lot, you become hot.

To summarise, we can now say that in the living cells of both plants and animals energy is released from the food with the help of oxygen, and carbon dioxide and water are formed as waste products. This method of energy release is known as **aerobic respiration**.

Anaerobic respiration

The next question is whether energy can be released from food without oxygen, i.e. **anaerobically**. You can find this out by using dried or baker's yeast. Looked at under the microscope it is seen to consist of a vast number of single cells. It is in fact a living fungus.

The principle of this experiment is to supply the yeast with glucose to act as food, but to deprive it of oxygen and see if energy and carbon dioxide are released.

Boil some 10% glucose solution to remove any oxygen dissolved in it, *cool* and pour into a small vacuum flask until about three-quarters full. Add a few grammes of yeast and pour a layer of liquid paraffin on top to prevent the mixture from coming in contact with atmospheric oxygen. Set up the rest of the apparatus as in Fig. 13:5. By taking the temperature at intervals over the next few days

130

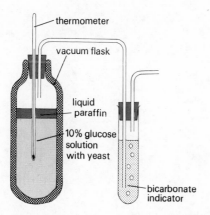

thermometer

vacuum flask

liquid paraffin

10% glucose solution with yeast

bicarbonate indicator

Fig. 13:5 Apparatus to test whether yeast liberates energy and carbon dioxide in the absence of oxygen.

you can find out if energy has been released, and by noting any change in the colour of the bicarbonate indicator you will discover if carbon dioxide has been formed.

By this single experiment you will not have proved that your result is due to a *living* process, or that any released energy comes from the glucose. To do this you must set up suitable controls using two similar sets of apparatus, except that in one the yeast must be dead (boil it up in the glucose solution first) and in the other no glucose must be added. Remember that in all three cases the temperature must be the same at the start.

From this experiment we should be able to deduce that yeast in the absence of oxygen does produce energy and does give off carbon dioxide. You will remember that with aerobic respiration the breakdown of the food takes place through a series of reactions, with the

Fig. 13:6 High power photomicrograph of budding yeast cells.

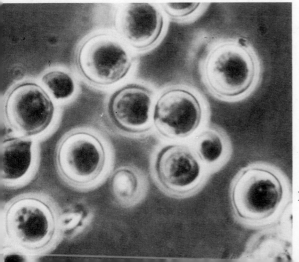

formation of various intermediate products. Some of the earlier reactions do, in fact, take place without oxygen and are therefore anaerobic; it is only the later ones that need oxygen. When yeast respires in the absence of oxygen the breakdown only goes as far as these anaerobic reactions, so much less energy is released and the glucose is turned mainly into ethanol—a substance which has still a lot of energy trapped in it. Anaerobic respiration in yeast can be summarised by this equation:

$$\underset{\substack{\text{(containing} \\ \text{trapped} \\ \text{energy)}}}{\underset{\text{glucose}}{C_6H_{12}O_6}} \xrightarrow{\text{enzymes}} 2CO_2 + \underset{\text{ethanol}}{2C_2H_5OH} + \underset{\text{(released)}}{\text{energy}}$$

Although the amount of energy released anaerobically is small compared with the aerobic method, it is enough to keep organisms like yeast and some bacteria active and most plants can survive for short periods by this means if the oxygen is temporarily cut off.

The formation of ethanol from sugar by yeast is known as **fermentation**. It is the basis of brewing and wine making. To make beer, yeast is added to malt (crushed germinating barley in water) and the ethanol is formed from the sugar in the malt; hops are added to give it a bitter taste. A wine is formed when sweet juices, usually from fruits, are allowed to ferment. Natural yeasts occur on the skins of fruit, e.g. the bloom on grapes, so when grapes are crushed the sweet juices become mixed with the yeast and fermentation takes place.

Yeast is also used in bread-making to make the dough rise. When yeast with a little sugar solution is mixed with the dough (flour, water and a little salt), it respires anaerobically and the carbon dioxide given out, being a gas, 'blows up' the bread.

You should now be able to explain why:
1. People who make home-made wine must not cork the bottles too soon.
2. A baker must allow time for the dough to 'rise' in a warm place before he puts it in the hot oven.
3. Holes are sometimes found inside a loaf.

131

Fig. 13:7 Finish of the 100m men's final, Olympic Games 1972. The oxygen debt built up during the race is repaid during the deep breathing that follows.

Muscles and energy

When we undertake strenuous physical exercise, it is not long before our muscles begin to ache and we become fatigued. One way in which physiologists have been able to study this process is by using a special kind of bicycle (Fig. 13:8). It is stationary and so enables the researcher to conduct tests on a volunteer more easily. The amount of effort needed to keep up a constant 'speed' can be increased by the use of electromagnetic brakes, so a state of near exhaustion can be reached quickly.

When blood samples are taken from a volunteer before and after a period of vigorous exercise, it has been found that a substance called lactic acid accumulates in the blood. But during a subsequent period of rest the amount of lactic acid is gradually reduced to its original level.

Lactic acid is an intermediate product in glucose breakdown. It is produced *without* the

Fig. 13:8 Research bicycle used to study the physiology of vigorous muscular action.

132

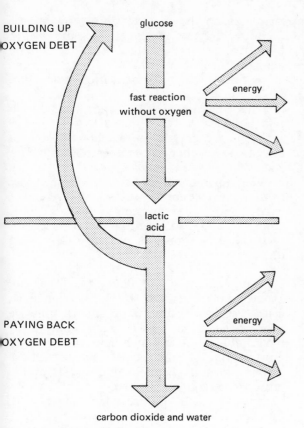

BUILDING UP
OXYGEN DEBT

glucose

fast reaction
without oxygen

energy

lactic
acid

PAYING BACK
OXYGEN DEBT

energy

carbon dioxide and water

Fig. 13:9 Diagram summarising how energy is released during and after muscular exercise.

help of oxygen, and some energy is released in the process. But when there is plenty of oxygen the reaction goes further; some of the lactic acid is completely oxidised to carbon dioxide and water with the liberation of more energy, and the remainder is reconverted into glucose (Fig.13:9).

During a 100 m race a well-trained athlete can hold his breath all the time—it is not until afterwards that he pants. In this case, the muscles are using the energy released during the anaerobic breakdown of glucose. It is not until afterwards that the athlete obtains the oxygen needed in order to remove the lactic acid. Therefore, when we undertake strenuous exercise we build up what is called an **oxygen debt** which has to be repaid later. In a longer race athletes have to breathe all the time, so some lactic acid is removed while they are running, and they can go on for longer before becoming exhausted. The presence of lactic acid in the blood is the main cause of muscle fatigue, but if the body is rested for long enough the tiredness goes.

14

Gaseous exchange

Respiration and the size of organisms

Every organism must have a surface through which oxygen and carbon dioxide can pass and also a means of taking these substances to or from the cells. The methods used vary according to the size of the animal and whether it lives in water or on land. Aquatic organisms utilise the oxygen which is dissolved in the water, terrestrial ones obtain it from the air.

Because microscopic, single-celled organisms such as amoeba are so small, the surface area is large compared with the volume, and oxygen can easily diffuse from the water through the surface into the cytoplasm. Simi-

larly, carbon dioxide can diffuse out in the opposite direction. As organisms increase in bulk the ratio of surface area to volume changes. You can see that this is so by calculating this ratio in cubes of various sizes (Fig. 14:1).

Thus increase in size brings with it two difficulties:
1. More oxygen is required, but the surface for gaseous exchange becomes relatively less and is therefore inadequate.
2. Many of the cells are some distance away from the surface and so are less likely to receive oxygen because it may be used up on the way. Thus larger animals must somehow increase their surface for gaseous exchange and have a means of taking the oxygen to the cells deep in the body (Fig. 14:2).

Simple animals like flatworms solve the problems by being flat. In this way they have a large surface/volume ratio and all cells are near the surface. Most animals, however, have a blood system which helps in two ways. First, through being in close contact with the respiratory surface, blood increases the efficiency of gaseous exchange by quickly removing oxygen which has diffused in, so increasing the rate of diffusion (p. 121). Secondly, it acts as a transport system, taking the oxygen to the cells.

Insects, as we saw in Chapter 6, have a unique method of bringing oxygen directly to the cells by means of an elaborate system of branching tubes, the tracheal system. This overcomes the necessity for having a blood system to transport the oxygen. It is a very successful method for small animals but is impracticable for large ones.

For animals like earthworms which are relatively sluggish, and so need comparatively little energy, the skin, with its special blood vessels, provides enough surface for gaseous exchange. But with larger, active animals this is not enough, so they have special respiratory organs which provide a greatly increased surface in contact with the water or air; these organs are called **gills** and **lungs** respectively.

The gill filaments of fish provide a very great surface area compared with the volume of the fish and each filament contains blood vessels very near the surface (p. 84). What is more, the water containing the oxygen is actively pumped over the filaments making the

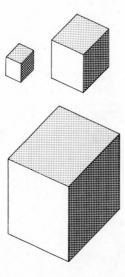

Fig. 14:1 Calculate the change in the ratio of surface area to volume of these cubes which have sides in the ratio of 1:2:4 respectively.

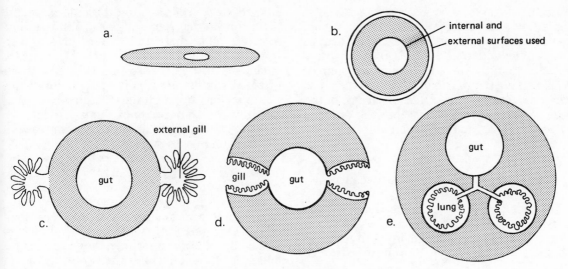

Fig. 14:2 Schematic diagrams of sections through various animals showing how the surface area has been increased for gaseous exchange. a) A flat body e.g. flatworms. b) A hollow body e.g. *Hydra*. c) Having external gills e.g. tadpole. d) Having internal gills e.g. fish. e) Having lungs e.g. mammal.

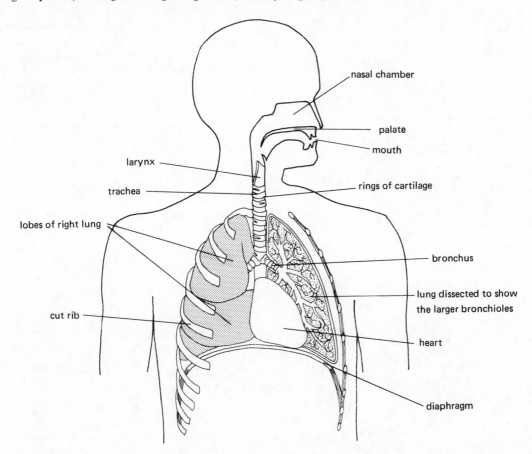

Fig. 14:3 Respiratory organs of man (pleurae not shown).

diffusion of gases even more efficient. Air-breathing vertebrates possess lungs which provide *within* the body a large surface area in close contact with blood vessels. The air enters the lungs from the outside through a system of tubes. During the course of evolution there has been a tendency for the internal surface of the lungs to increase, thus enabling the animals to become larger and more active.

Breathing organs in man

Fig. 14.3 shows the main structures concerned with respiration.

The two lungs, which are lobed, are spongy elastic organs filling up most of the chest cavity. They look solid, but in reality consist of a mass of tubes called **bronchioles** which

branch more and more and become finer and finer until they end up as blind-ending **alveoli**. It is these millions of alveoli that provide the great surface area for gaseous exchange, as each is surrounded by a network of blood capillaries. It is possible to inject the blood vessels of the lung of a mammal after death with a coloured latex. This enables you to see the network of capillaries associated with the alveoli (Fig. 14:5).

The lungs are in direct contact with the atmosphere through the nostrils, nasal cavity, larynx, trachea (wind pipe) and the two bronchi. Each lung is enclosed by two transparent membranes, the **pleurae**. The inner one covers the lung itself and the outer is attached to the chest wall. In between there is a little fluid which allows the two membranes to slide over one another easily when the lungs move. If the pleurae are damaged by bruising or disease they secrete more fluid into the cavity between them, causing pain and difficulty in breathing. This condition is known as **pleurisy**.

The chest cavity is air-tight and separated from the abdomen by a dome-shaped structure, the **diaphragm**. The sternum, ribs and backbone, together with the muscles of the chest wall, provide a tough box which surrounds the lungs and heart and affords them essential protection.

Fig. 14:6 Model thorax to demonstrate the action of the diaphragm during breathing.

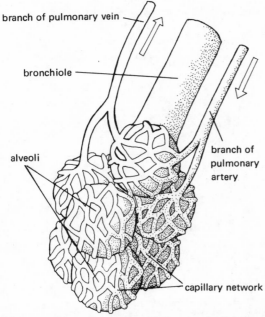

branch of pulmonary vein

bronchiole

alveoli

branch of pulmonary artery

capillary network

Fig. 14:4 Diagram of alveoli showing blood supply.

Fig. 14:5 High power photomicrograph of injected lung showing blood capillaries.

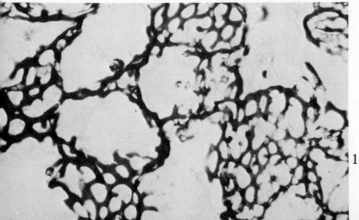

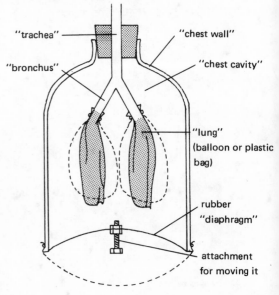

"trachea"

"chest wall"

"bronchus"

"chest cavity"

"lung" (balloon or plastic bag)

rubber "diaphragm"

attachment for moving it

136

The breathing mechanism

The basic mechanism can be explained through a model (Fig. 14:6). Air enters the lungs (in this case balloons) only if the pressure inside them is momentarily reduced to below that of the atmosphere. As the lungs are thin-walled, this can be brought about by reducing the pressure in the space around them so that they expand. To reduce this pressure, the volume of the space surrounding the lungs must be increased. This can be achieved in the model by pulling down the rubber sheet. In man it happens in two ways: by lowering the diaphragm and by moving the ribs upwards and outwards. The model illustrates only the movement of the diaphragm.

The diaphragm consists of a central non-elastic membrane and an outer ring of muscle which curves downwards to join the body wall (Fig. 14:7). When you take in a breath (inspiration), the radially arranged muscles of the diaphragm contract causing the membrane to be lowered. (The liver and stomach are pushed down slightly as a result.) At the same time, the rib cage is moved upwards and outwards by the contraction of the external intercostal muscles which are arranged obliquely between the ribs (Fig. 14:8).

Breathing out (expiration) is a more passive process, because it involves the relaxation of the diaphragm and external intercostal muscles. The lungs, being elastic, shrink when the pressure around them increases. In addition, there are internal intercostal muscles running at right angles to the external intercostals which contract and help to lower the

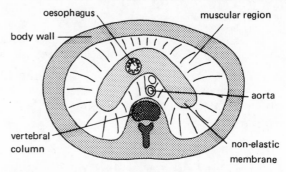

Fig. 14:7 Human diaphragm viewed from below.

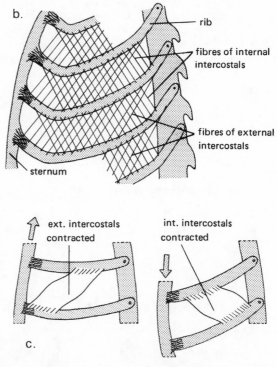

Fig. 14:8 Diagram to show the mechanism of breathing: a) Lateral view of thorax: left—breathing in, right—breathing out. b) The position of the external and internal intercostal muscles. c) The action of the intercostal muscles.

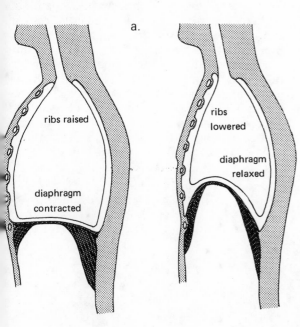

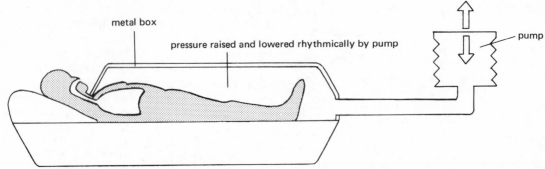

Fig. 14:9 Diagram showing the principle of the 'iron lung'.

rib cage. During expiration, abdominal muscles are also used to push the liver and stomach up against the diaphragm, thus forcing it further into the chest cavity.

The 'iron lung'

In cases of diseases which affect the functioning of the respiratory muscles (as sometimes happens with poliomyelitis) the patient's breathing movements have to be brought about artificially. This is done through an 'iron lung'. The patient is completely enclosed in a large metal box, except for his head, and the air pressure in the box is lowered and raised rhythmically. This has the effect of increasing and decreasing the volume of the whole chest. Thus when the pressure is reduced around the patient, the chest and the lungs expand and so air enters the lungs. When the pressure is increased around the patient air is expelled (Fig.14:9).

Artificial respiration

In cases where a person's breathing has stopped, for example through a drowning accident, the only hope of resuscitation is through the immediate application of artificial respiration. The patient's head is pressed well back to keep the trachea open; the mouth is opened and air is blown directly into the lungs from the rescuer's mouth. The patient's nostrils must be kept closed to prevent air from escaping (Fig. 14:10).

Control of the rate of breathing

Although our breathing movements are controlled by automatic nervous responses, the breathing rate varies according to circumstances. For example, when we take vigorous exercise, more of our food reserves are oxidised and so more carbon dioxide is produced. The latter alters the composition of the blood (it becomes more acid), and sensitive cells in the hind brain immediately react to this by increasing the rate at which nervous stimuli are sent to the intercostal and diaphragm muscles so that our breaths become quicker and deeper. The extra carbon dioxide is therefore quickly removed, the acidity of the blood decreases, and so the rate of breathing automatically slows down to its normal rhythm.

You could investigate this as follows:

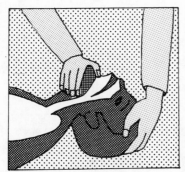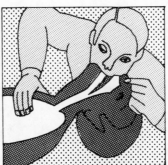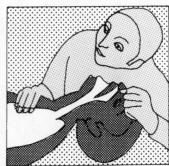

Fig. 14:10 The mouth-to-mouth method of applying artificial respiration (see text).

First count the number of breaths taken per minute when you are sitting down normally. From the data of the whole class prepare a block graph to show the average rate for the class and the range of variation.

The class should now decide on a set amount of vigorous exercise which is within the capabilities of everybody! (e.g. sprinting, press-ups, etc.).

Now, working in pairs, one should carry out the exercise. Immediately this has finished, the other should record the number of breaths his partner takes each half-minute until the rate becomes steady. Repeat the procedure with the roles reversed.

You should now plot a graph showing how your breathing rate altered during each half-minute following exercise.

Do you find some members of the class recover more quickly than others? Does the rate of breathing in different individuals vary a great deal with a similar amount of exercise? Try to suggest possible reasons for any differences.

Lung capacity

Another factor concerned with respiratory efficiency is lung capacity.

Measure the amount of air in your lungs by taking a very deep breath and blowing out as much air as possible under water using the apparatus shown in Fig. 14:11. This volume is known as the **vital capacity** of the lungs.

When the whole class has done this draw up a table showing the vital capacity of each member in one column and the normal rate of breathing in another. Can you discover from the more extreme cases whether there appears to be a connection between having a high or low vital capacity and having a high or low rate of breathing? Other factors can also be compared with these two measurements, such as chest expansion, and even athletic ability. Have good sprinters and long-distance runners in your class rather similar characteristics in these respects?

When you have breathed out all the air you can there is still about $1\frac{1}{2}$ litres of air locked up inside the alveoli so you can now work out the **total capacity** of your lungs.

Gaseous exchange

Because oxygen is being used up by the body and carbon dioxide given out, we would expect certain differences between inspired and expired air. One way of investigating this is to use a special capillary tube or J tube (Fig. 14:12). This enables you to measure the

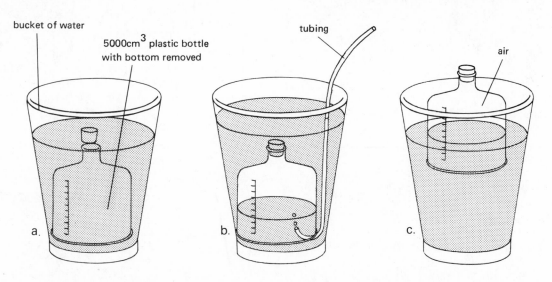

Fig. 14:11 Apparatus for measuring the vital capacity of the lungs: a) The bung is removed to fill the bottle and is then replaced. b) The bottle is held loosely to prevent tipping while air is exhaled through the tube. c) The volume of air is measured after levelling.

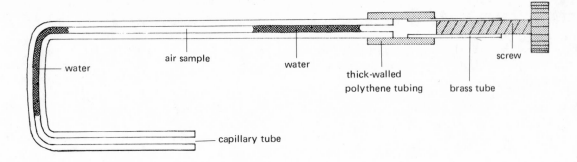

water — | air sample | water | thick-walled polythene tubing | brass tube | screw — capillary tube

Fig. 14:12 J tube.

amount of oxygen and carbon dioxide in a sample of air. It requires a little practice to operate it, so it is best to sample atmospheric air first (the air we breathe in). It is useful to work in pairs.

1. Turn the screw clockwise as far as it will go. Dip the end of the tube into water and turn the screw until a column of water 4 or 5 cm long has entered.
2. Lift the tube out of the water and draw in about 10 cm of air. Then immerse the end in the water again and draw in another 4 or 5 cm of water.
3. Adjust the screw so that the column of air lies along the long straight side of the J tube. The temperature of all the air samples you are going to measure must be constant (why?) so leave the tube in a sink full of water for a minute. Now measure the length of the air column under water with a ruler, keeping your fingers as far away from the column as possible. (Why?)
4. Turn the screw until the sample is within 2–3 mm from the end of the tube (no further!) and draw in about 5 cm of concentrated potassium hydroxide solution carefully. This will dissolve any carbon dioxide with which it comes in contact. Draw the potassium hydroxide backwards and forwards several times over the first part of the tube so that a thin film of it comes in contact with the column of air. Any carbon dioxide in the bubble should now be absorbed and the bubble will be reduced in length.
5. Expel the potassium hydroxide carefully, making sure the air sample remains in the tube. Draw the air back again into the long straight part of the tube, immerse in water

for a minute and then measure the length of the air column.
6. Repeat procedures 4 and 5, but this time draw in some alkaline pyrogallol which will absorb oxygen.
7. Before taking any further samples rinse the J tube thoroughly, first in dilute acid and then water. (Why?)
8. Now analyse a sample of exhaled air. It can be collected from a test tube as in Fig. 14:13. Take a fairly deep breath. Blow slowly down the rubber tube to remove the residual air inside it. While still blowing slowly place the end of the tube under the mouth of the test tube.

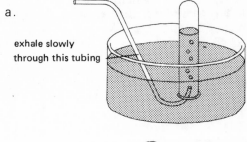

a.

exhale slowly through this tubing

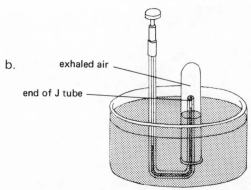

b.   exhaled air

end of J tube

Fig. 14:13 Method of obtaining a sample of exhaled air using a J tube. a) Breathing into a test tube of water. b) Inserting the J tube.

9. Take a sample of the exhaled air into the J tube and analyse as before.

NOTE: Potassium hydroxide and alkaline pyrogallol are both caustic and will burn the hands, so they should be handled with care.

Now calculate the percentage of carbon dioxide and oxygen as follows: Suppose, for example, the column of air was 10 cm long to begin with and the reading after treatment with potassium hydroxide was 9·6 cm and after treatment with alkaline pyrogallol 7·9 cm.

Reduction in volume with potassium hydroxide is $10·0 - 9·6 = 0·4$

Reduction in volume with alkaline pyrogallol is $9·6 - 7·9 = 1·7$

Therefore the percentage of carbon dioxide in the sample is

$$\frac{0·4}{10·0} \times 100 = 4\%$$

Similarly, the percentage of oxygen is

$$\frac{1·7}{10·0} \times 100 = 17\%$$

What conclusions can you draw from your results? Compare them with the following table which gives the average percentage composition:

	Atmospheric air	Alveolar air	Exhaled air
Nitrogen	79·01	80·7	79·5
Oxygen	20·96	13·8	16·4
Carbon dioxide	0·03	5·5	4·1

With quiet breathing we only exchange about 500 cm³ of air. Not more than about 350 cm³ of this actually reaches the alveoli because the rest fills the trachea and bronchi. Thus it follows that the amount of fresh air coming in has comparatively little effect on the composition of gases in the alveoli. Exercise, however, makes us breathe more quickly and deeply and more of the stale air is removed.

Evidence that the blood is involved in gaseous exchange can be provided by analysing the amount of the various gases in samples of blood entering and leaving the lungs. The following table gives the volume of each gas present in 100 cm³ of blood:

	Blood entering the lungs	Blood leaving the lungs	Change
Nitrogen	0·9 cm³	0·9 cm³	—
Oxygen	10·6 cm³	19·0 cm³	+8·4 cm³
Carbon dioxide	58·0 cm³	50·0 cm³	−8·0 cm³

The exchange of gases at the lung surface between the air and the blood takes place by diffusion (Fig. 14:14). This process is very efficient because the internal surface area provided by approximately 700 million alveoli in the two lungs totals over 70 m² and the capillary network provides an area of over 40 m². Compare these figures with the total surface area of the walls, floor and ceiling of a medium-sized room. In addition, the diffusion pathway, i.e. the distance between the air in the alveoli and the blood in the capillaries, is only about 1 μm. Also, because the alveolar walls are always moist, oxygen can dissolve more easily.

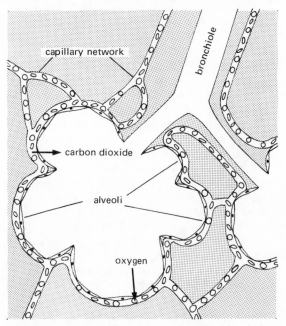

Fig. 14:14 Diagram of a section through the alveoli of the lung showing the diffusion pathway for gaseous exchange between lung and blood capillaries.

In the capillaries the oxygen is taken up by a red pigment, **haemoglobin**, contained within the red blood corpuscles. In this way blood can carry more oxygen than in simple solution.

The respiratory system and health

The respiratory system is open to the external environment via the nose and mouth, and is therefore a source of entry for pathogenic (disease-causing) organisms and atmospheric pollutants which might damage the delicate tissues of the lungs.

The system has its natural defences against these invaders. The nostrils and the hairs within them are coated with a slimy secretion, **mucus**. The hairs act as an efficient sieve for larger particles; after blowing your nose when you have been in a dusty or sooty atmosphere you will see plenty of evidence for this. However, if you breathe through your mouth you derive no benefit from this protection. In addition, particles may stick in the mucus secreted by cells which line the trachea and bronchioles. Most of these cells have cilia which are in constant motion, causing the mucus to pass slowly up the tubes to the back of the throat where it is swallowed unconsciously. The particles are carried with the mucus as on a conveyor belt, away from the lungs.

These defences are not always sufficient. People who are constantly inhaling gritty particles are liable to damage their lungs badly and develop various diseases such as **silicosis** and **asbestosis**. In these situations, the workers should wear masks to avoid breathing in the harmful particles.

Atmospheric pollutants also have a major effect on health. During the combustion of fuel for generating electricity vast quantities of smoke, grit, sulphur dioxide and other substances are passed into the atmosphere. In addition the exhaust fumes of cars add poisonous carbon monoxide and oxides of nitrogen and lead. All these substances may affect the respiratory system and lead to ill health.

All parts of the respiratory system may be attacked by bacteria or viruses. Usually the body defences are sufficient but, if not, the micro-organisms invade the tissues and cause inflammation and disease. If this occurs at the back of the throat (pharynx) a sore throat or **pharyngitis** develops, if in the larynx, **laryngitis**, which may cause you to lose your voice; infection of the trachea is known as **tracheitis** and of the bronchioles, **bronchitis**. The cells respond to irritation or an attack by secreting mucus, and if this obstructs any of the tubes it produces a reflex action which is a cough. If a lot of mucus is present in the bronchioles, a doctor by using a stethoscope can hear the 'bubbles and squeaks' and can diagnose bronchitis. If the organisms reach the air sacs and finer bronchioles, the attack causes them to become filled with fluid and they can no longer carry out gaseous exchange. This condition is called **pneumonia**. In serious cases patients are given oxygen to make up for this loss of respiratory surface area.

Smoking and health

One of the major irritants to the respiratory system is cigarette smoke. Its nicotine content stops the ciliary action of the lining cells and the irritation causes more mucus to be produced, so it has constantly to be coughed up. This results in 'smokers' cough' which is the first stage of chronic bronchitis.

Chronic bronchitis is a major health hazard and accounts for 30,000 deaths each year in Britain. Of every 8 men suffering from bronchitis, 7 are smokers. Chronic bronchitis is also brought on by other air pollutants such as the smoke from factory and domestic chimneys and the exhaust fumes from cars. Legislation has done much already to reduce smog (e.g. smokeless zones in cities), but exhaust fumes still remain a major menace.

Another effect of smoking is on the growth of the lungs. It has recently been shown that smoking by children prevents the normal growth of the lungs and permanently affects their capacity and efficiency. But the most serious effect of smoking is lung cancer. It is caused by substances in tobacco smoke called **carcinogens**. Unfortunately this is a disease which usually develops twenty or more years after the smoking habit has been acquired and then it is often too late to do much about it. Cases of lung cancer are increasing all the time. In the U.S.A. 2,500 died of it in 1930; in 1967 the number had increased to 50,000. In Britain, 27,000 men and 5,500 women died of it in 1968, about half of them before the age

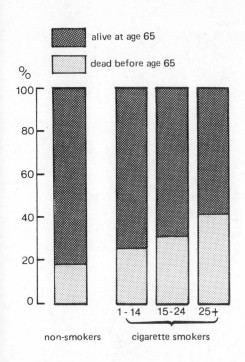

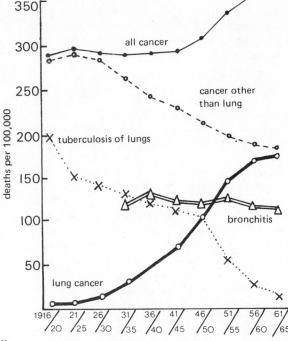

Fig. 14:15 (left) Proportion of men aged 35 who will die before the age of 65 according to their smoking habits. (right) Death rates from lung cancer, other forms of cancer, tuberculosis of the lung and bronchitis in men aged 45–64 from 1916–1965.

of 65. The only group of people in which lung cancer deaths have gone down in recent years has been the doctors: they know the dangers!

The nicotine in cigarette smoke is not only a drug which results in dependence but also has important physiological effects especially on the circulatory system. Cigarette smoking is known to be an important factor in bringing about coronary thrombosis (heart attack) and mothers who smoke a lot not only tend to have smaller babies, but the risk of spontaneous abortion and still birth is increased. It is estimated that 1,500 babies in the United Kingdom were lost in 1970–71 through their mothers' smoking habits during pregnancy. Certain forms of heart disease in babies are also significantly more common when their mothers are smokers. Another danger is that cigarette smoke contains carbon monoxide. This highly poisonous gas interferes with the capacity of the blood to carry oxygen. In heavy smokers it may lead to a loss of oxygen-carrying capacity of up to 10%.

The seriousness of the effect of smoking can be gauged by the fact that 30–35,000 people in Britain die every year from lung cancer and a similar number from bronchitis, coronary thrombosis and other causes directly attributable to cigarette smoking. Put in another way—people are being killed by cigarettes in Britain alone at the rate of 1 every 10 minutes. This is 4 times as many as by all road accidents. Smoking 20 cigarettes a day increases by 20 times the chance of contracting lung cancer, compared with the non-smoker. One in 8 people who smoke 40 cigarettes a day contract the disease.

With these facts available you may well wonder why people smoke at all. The reason is basically because nicotine is a drug of dependence. Nearly all smokers will tell you that they started because they wanted to appear tough or grown up, to be sociable or, more often, just to see what it was like. They became dependent long before they realised and then found it very difficult, if not impossible, to drop the habit.

Respiration in plants

When we investigated aerobic respiration in the last chapter we used both animal and plant material according to convenience. This was done because plant and animal respiration are basically the same. It is only in the means by which oxygen reaches the cytoplasm that differences occur.

Respiration takes place in every living cell, so there must be a means of oxygen entry into the plant, a method of transporting it to every cell and a means of removing the products of respiration (carbon dioxide and water).

Gaseous exchange takes place through vast numbers of microscopic holes in the leaves and young stems called **stomata**. They are capable of opening and closing. Their structure will be considered on p. 152. On the surface of woody stems and roots there are different openings called **lenticels** which serve the same purpose.

Inside the plant these openings lead to a series of spaces between the cells which form a continuous network all over the plant. The spaces are very large in the leaves, much smaller in other parts of the plant. The air spaces are lined with water and the oxygen in the air spaces dissolves in this and passes through the porous cell walls into the cytoplasm where the sugar is broken down into carbon dioxide and water with the liberation of the energy. The carbon dioxide passes out into the air spaces by a similar method.

The whole system works by diffusion; as the oxygen is used up by the cells a gradient develops between the cells and the air in the spaces, and similarly between the air in the spaces and the air outside the stomata and lenticels, so oxygen passes in. In the same way, as more carbon dioxide is given out by the cells a gradient occurs in the reverse direction and it passes out.

Most plants can aerate their roots by taking in the oxygen through the lenticels or through the surface of their root hairs (as their walls are very thin). They obtain this oxygen from the air spaces between the soil particles. But plants which have their roots in very wet places, such as ponds or marshes, are unable to do this. They are adapted to these water-logged conditions by having much larger air spaces which

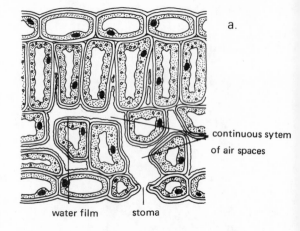

continuous sytem of air spaces

water film   stoma

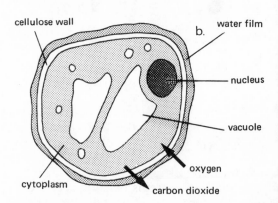

cellulose wall   water film

b.

nucleus

vacuole

oxygen

cytoplasm   carbon dioxide

Fig. 14:16 The leaf as a respiratory organ. a) A transverse section showing system of air spaces. b) A single leaf cell showing diffusion pathways.

connect the stems with the roots, making diffusion from the upper parts much more efficient. The most usual adaptation is to have a hollow stem. Next time you are by a pond or marsh cut the stems of some of the plants which are growing there and see how many are hollow compared with a similar number of species of plants growing in normal soil. The problem of air transport is more difficult for trees and not many survive with their roots permanently in water. An exception is the mangrove tree of the tropics which sends up aerial roots above the surface and takes in oxygen that way.

15

Nutrition in green plants

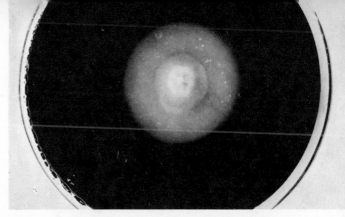

Fig. 15:1 Starch digestion by a saprophyte. The fungus (centre) is growing on agar containing starch. When flooded with iodine the plate turns black except where the fungus is growing.

Methods of nutrition

Before considering nutrition in green plants in some detail, it is useful to summarise the three main methods of nutrition used by living organisms as there is a close relationship between them.

1. Holophytic nutrition

This is the method used by green plants. It involves the building up of carbohydrates from simple chemical substances by a process called **photosynthesis,** and also the formation of fats and proteins by further processes. This is said to be an **autotrophic** or self-feeding method because in this way they *make their own food.* (All other organisms are dependent for their nutrition on those substances that green plants have produced; they are therefore said to be **heterotrophic** because they have no means of building up their own food from simple substances.)

2. Holozoic nutrition

This is the method characteristic of most animals. It occurs in three stages: a) **ingestion**—the taking in of the food at a particular place (in most animals the mouth); b) **digestion**—the processing of the complex food molecules by means of digestive enzymes until the products are soluble and have small enough molecules to be absorbed; c) **absorption**—the extraction of these products from the digestive system into the blood, from which it can be taken up by the cells and utilised. In the simpler animals, which have no blood system, the digested products diffuse directly into the surrounding cytoplasm.

3. Saprophytic nutrition

This is the method used by plants which have no chlorophyll, such as fungi (p. 41) and most bacteria (p. 189). Like all plants, saprophytes have no mouths, so there is *no* process of ingestion. Instead, enzymes are secreted by the cytoplasm through the cell wall on to the food, a process called **external digestion**. The soluble products are then absorbed and utilised. Saprophytic fungi and bacteria thus feed on non-living organic matter and by their feeding process cause it to decay. A few simple animals which have no mouths also absorb their food through their surfaces in a similar way, for which the term **saprozoic** is used.

In contrast to the saprophytes which feed on non-living material, there are **parasites** which are dependent on other *living* organisms for their food. They live either on their hosts (**ectoparasites**), e.g. fleas and lice, or inside them (**endoparasites**), e.g. tapeworms, often harming them in the process. Parasites may either feed directly on the living cytoplasm, e.g. the malarial parasite which feeds on red blood corpuscles (p. 71), or on the food of the host, e.g. gut parasites such as tapeworms (p. 36). The actual process of nutrition, however, is not basically different from either holozoic or saprophytic nutrition because the food is either ingested as in typical animals, or absorbed through the surface as in saprophytes and saprozoic organisms. Thus it is better to think of parasitism as a different mode of life rather than a different method of nutrition.

HOLOPHYTIC NUTRITION

Every green plant is composed of living protoplasm and the non-living substances that the protoplasm has produced. Many of these substances are complex, but all are synthesised from relatively simple chemicals which the plant finds in its immediate surroundings. Plant nutrition involves the taking in of these nutrients and their synthesis into more complex foods.

The substances which are synthesised are used for two main purposes: the growth of the plant, including the formation of new individuals (reproduction), and respiration, the means by which the plant obtains its energy for all its metabolic activities.

Photosynthesis

This is the basic process of plant nutrition. If green plants did not photosynthesise we could not exist, as all the food we eat, whether animal or vegetable, has its origins in this process. Because of the great importance of photosynthesis we will investigate it in some detail.

If carbohydrates are made when green plants photosynthesise we should be able to test for the presence of these substances. Let us start with starch.

Testing for starch

A potato is full of starch: pipette some iodine dissolved in potassium iodide on to its cut surface. Iodine turns starch blue-black. Look at the treated potato under a lens and notice how it is really speckled all over. Each speck is a solid starch grain.

Can starch be detected in green leaves?

It is not so easy to test for starch in a green leaf because the green colour prevents the starch from being seen even after treatment with iodine. But you can do it by first removing the chlorophyll with ethanol. Proceed as follows:

With a cork borer punch out some discs from a thin green leaf which has been in good light for some hours, e.g. nasturtium, lime or lilac.

Plunge the discs into a beaker of boiling water for half a minute and transfer them with forceps to a test tube containing a little ethanol.

Remove the bunsen from under the water bath before placing the test tube in the water. This is because ethanol is very inflammable and the contents of the tube might catch alight. The ethanol will boil quickly as its boiling point is below that of water.

When the ethanol has dissolved out all the green chlorophyll, remove the discs with forceps, dip them in water, place them in a clean test tube and just cover with iodine solution. After a few minutes, pour off the iodine, wash in cold water and note whether the blue-black colour is present in the leaves. If it is, we can conclude that starch is present.

Is light necessary for starch formation?

Now that you know how to test for starch in a leaf, you can find out if light energy is essential for its formation. It would be possible to adapt the last experiment, by taking discs from leaves which have been kept in the dark for 24 hours, and comparing them with other discs exposed to light for some time. Another way is to experiment on a single leaf and expose only part of it to light.

Make a sleeve of black paper with a pattern cut out of it as in Fig. 15:2. Attach it to the leaf of a shrub or tree where it will receive good light, leave overnight and for as long as possible the next day before testing for starch. Does the distribution of starch correspond to the part of the leaf which was exposed to the light? (Instead of black paper it is possible to use a contrasting photographic negative of a simple subject. This should produce a recognisable picture on the leaf when you 'develop' it in iodine.)

What other factors could be involved in starch formation? As starch is found in green leaves exposed to light, it is possible that the chlorophyll which produces the greenness plays a part in its formation.

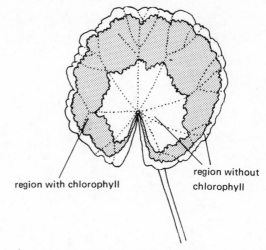

Fig. 15:3 A variegated leaf—used to determine whether chlorophyll is needed for starch formation.

region with chlorophyll

region without chlorophyll

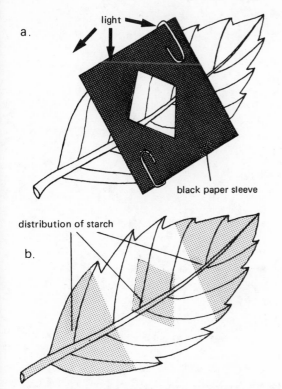

a.

light

black paper sleeve

distribution of starch

b.

Fig. 15:2 a) Method of demonstrating the effect of light on starch formation in a leaf. b) The same leaf after testing for starch.

Is chlorophyll necessary for starch formation?

To investigate this you can use a leaf which has its own built-in control. Certain plants have leaves which are partly green and partly white (or yellow), e.g. some forms of geranium. Remove one of these **variegated** leaves which has been exposed to light for some hours. Draw the pattern to show the green and white parts of the leaf, extract the chlorophyll in ethanol as in the previous experiment, and then test for starch. Does the pattern of the starch correspond with the green part of the leaf?

From these experiments it can be shown that light energy and chlorophyll are both necessary for starch formation, but what is the starch made from? The starch molecule is composed of a variable number of units formed from glucose, each with the basic formula $C_6H_{10}O_5$, so obviously substances containing carbon, hydrogen and oxygen must be involved. A readily available carbon com-

pound is carbon dioxide in the air. We know that in respiration carbon dioxide is given out by plants. Is it also used in the formation of starch when light and chlorophyll are present?

Is carbon dioxide taken in when green leaves are exposed to light?

You can find out if this occurs by carrying out an experiment involving bicarbonate indicator, the colour of which varies according to the concentration of carbon dioxide: no carbon dioxide—reddish purple; very little carbon dioxide, e.g. when the indicator is exposed to air—orange red; greater concentrations of carbon dioxide—yellow.

We could use this indicator to see if any changes occur in the carbon dioxide content of the atmosphere when green leaves are illuminated.

1. Wash out four boiling tubes thoroughly in tap water, rinse with distilled water and finally with a little bicarbonate indicator.
2. Put 2 cm³ of bicarbonate indicator in each and stopper at once (why?).
3. Wrap black paper around Nos. 3 and 4 to keep out the light.
4. Place similar green leaves which have just been removed from a healthy plant in sunlight in Nos. 2 and 4 in such a way that they do

147

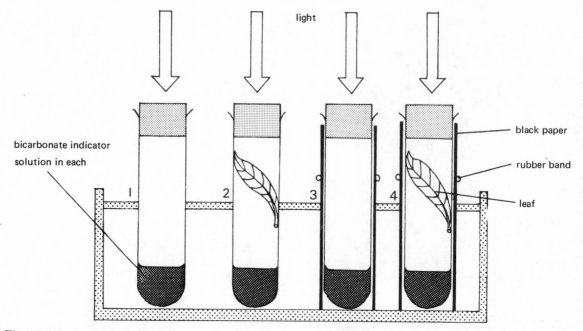

Fig. 15:4 Apparatus to determine if any changes occur in the carbon dioxide content of the atmosphere when green leaves are illuminated.

not slip into the indicator; do not put a leaf into Nos. 1 and 3.

5. Place all four tubes in a place where light from the window or bench lamp can illuminate them equally.

6. Shake each *gently* every five minutes and after half an hour compare the colours of the indicator in each by holding them against a white background. Compare tube 1 with 2, 3 with 4, 1 with 3 and 2 with 4. What can you deduce from each of these comparisons? What is your final conclusion of the effect of leaves in light and darkness on the carbon dioxide content of the atmosphere?

Green plants in the *dark* give off carbon dioxide as a result of *respiration*, but you should have discovered from the last experiment that when green leaves are illuminated no carbon dioxide is given off. It is therefore reasonable to suppose that it is used up again, but is it used in the making of the starch or for some other purpose? We could try to find out by designing an experiment to see if carbon dioxide is essential for starch formation.

Is carbon dioxide necessary for starch formation?

Fit up the apparatus as shown in Fig. 15:5. Put some caustic potash solution in one jar to absorb the carbon dioxide, and potassium bicarbonate solution in the other to ensure that a good supply of carbon dioxide is in the air above. Put in each tube a nasturtium or other suitable leaf which has previously been kept in the dark to remove any starch. Absence of starch can be proved by taking out a disc from each leaf with a cork borer and testing as before. Keep the leaves strongly illuminated for several hours, preferably all day, and then test for starch in each. From your results deduce whether carbon dioxide is necessary for starch formation.

We described on p. 128 how scientists have used radioactive isotopes to 'label' certain atoms so that they could be traced from molecule to molecule. By a similar technique, using radioactive carbon in the carbon dioxide supplied to plants, it has been confirmed that it is the carbon atoms in *carbon dioxide* which are passed into the starch molecule.

148

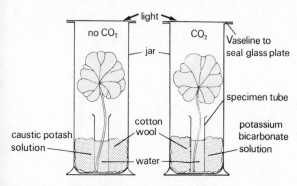

Fig. 15:5 Experiment to find out whether carbon dioxide is necessary for starch formation.

From these investigations we should now be able to conclude that light, chlorophyll and carbon dioxide are all needed in the making of starch. The next question is whether any products are given off during starch formation. When considering respiration we saw that oxygen was used up and carbon dioxide given out. In photosynthesis we have seen that carbon dioxide is used up. Is oxygen put back into the air?

Is oxygen given off by green plants in the light?

It is not easy to investigate this on a land plant as it is already surrounded by air of which about one-fifth is oxygen, but by using a water plant any oxygen given off would appear as bubbles, as very little oxygen dissolves in water and water soon becomes saturated.

If you observe some Canadian pondweed, *Elodea*, in a beaker of water when it is in bright light, you will notice that bubbles of gas are given off from the cut ends of the stems. By collecting these in sufficient quantity you could test whether this gas is oxygen or not.

Set up the apparatus as shown in Fig. 15:6. Use plenty of sprigs of *Elodea* and cut the ends with scissors before putting them in the beaker. Why should the funnel be kept off the bottom of the beaker? Use a bench lamp to illuminate the plants strongly. If the apparatus can be left for several days you will have more gas to analyse.

When enough gas has collected, use a J tube to find the percentage of oxygen in the sample. Proceed as you did for analysing air (p. 140). To obtain the sample, first draw in a

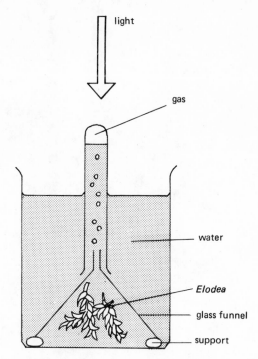

Fig. 15:6 Apparatus to determine whether oxygen is given off by green plants in the light.

little water, then push the end of the J tube under the test tube until it reaches the gas, draw in a column of gas about 10 cm long, withdraw the tube until its end is under water again and draw in a little more water to seal off the gas. Place the tube under water in a sink for a minute and measure the column of gas accurately. Now proceed as before, first using potassium hydroxide to absorb any carbon dioxide and then alkaline pyrogallol to absorb the oxygen.

Calculate the percentage of oxygen in the gas. Is the percentage greater than that of air?

Does the intensity of light affect the rate of photosynthesis?

To find out, you could modify the last experiment by taking a single sprig of *Elodea*, counting the bubbles given out by the cut end, and then, by altering the light conditions, see if the speed of bubbling changed. Proceed as follows:

Fit up the apparatus as in Fig. 15:7. Shine a strong lamp on to the *Elodea* horizontally.

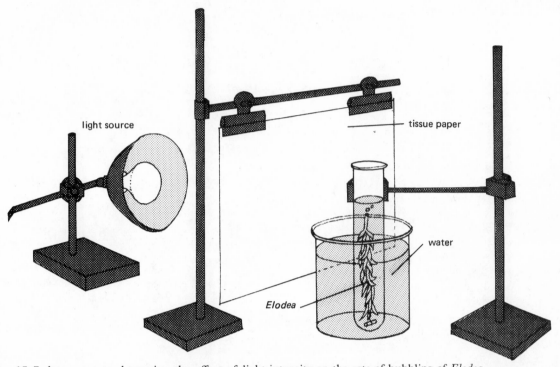

Fig. 15:7 Apparatus to determine the effect of light intensity on the rate of bubbling of *Elodea*.

Why is the tube immersed in a beaker of water? Keep the bench lamp the same distance from the *Elodea* during the whole of the experiment. Why? Leave for 3 minutes, then count the bubbles for 3 successive minutes and take the average. Now alter the light intensity by putting a sheet of tissue paper in front of the lamp. Leave for another 3 minutes and then take three more counts, add another sheet and do the same until you have a series of average rates to compare with changes in intensity. Make a graph showing rate of bubbling against the number of sheets of paper. If a light meter is available, you could measure the light intensity accurately. Place the meter level with the *Elodea* and, pointing it towards the light source, take a reading for each intensity. Then plot the number of bubbles against light intensity. What can you deduce from your results?

In the experiments so far, we have assumed that starch is formed as a result of photosynthesis, but a molecule of starch consists of many glucose molecules connected together. It would therefore be a likely hypothesis that during photosynthesis glucose is formed first and then changed into starch.

Glucose belongs to a group of substances called reducing sugars. Their presence can be demonstrated using the following test:

Testing for reducing sugars

Add 5 cm³ of Benedict's solution to each of two tubes. Add a little powdered glucose to one and use the other as a control. Heat both to boiling point. Notice how the glucose causes the blue colour to change until finally a brick-red precipitate is formed. This precipitate still clings to the test tube after you have washed the contents down the sink. (It can be cleaned off with dilute hydrochloric acid.) The colour change occurs because the copper compound in the Benedict's solution is reduced by the glucose to red cuprous oxide. Is there any change in the control tube?

Now take another clean boiling tube, but this time add only a few grains of glucose to the Benedict's solution. Boil as before. Is there any change of colour?

150

When you test for reducing sugar in leaves you will be dealing with very small quantities of sugar, so remember that *any* change in colour is a good test for its presence. The colour varies with the quantity of sugar present.

Is reducing sugar present in leaves?

Take three kinds of leaves which have been exposed to light for some hours: iris, geranium and lilac would be a good selection. Cut out four discs of iris with a cork borer and grind them up with a pinch of sand and a little water; filter and test the filtrate for reducing sugar with Benedict's solution. Repeat for the other two leaves and compare the results. Which leaves contained reducing sugar?

Your results will have shown you that some leaves do form sugar, but others apparently do not. A possible explanation of this result could be that in those which showed no reducing sugar, it was in fact formed, but it was converted into starch immediately.

To investigate this possibility you could see if the plants which showed no reducing sugar in the last experiment are able to produce starch if kept in the dark and supplied with glucose.

Can leaves turn glucose into starch in the dark?

Using plants which showed no reducing sugar in their leaves, keep them in the dark for two days. Punch out a number of discs. Test one to make sure that no starch is present.

Take two Petri dishes, place 5% glucose solution in one and water in the other to act as a control. Float some discs on each with their undersurfaces in contact with the liquid. Keep in the dark for several days and then test for starch.

This experiment should clearly demonstrate that starch can be made from glucose sugar by the leaf and that this reaction does not need light. In fact, research has shown that glucose is the primary carbohydrate formed in photosynthesis, but in many plants it is immediately converted to starch. The conversion of sugar into starch is advantageous to the plant as sugar is soluble in water; if it became too concentrated, it would act osmotically and draw water into the cell. Starch by contrast is insoluble so there is no danger of this. However, as we have seen from our experiment, some plants do have some sugar in their leaves when they are exposed to light.

How does photosynthesis take place?

So far we have seen that light energy, chlorophyll and carbon dioxide are needed for photosynthesis to take place, and oxygen is released. Looking at the formula for glucose sugar, $C_6H_{12}O_6$, it can be seen that some hydrogen atoms have to come from somewhere. It has been shown by using radioactive isotopes that these come from water and that the oxygen given off also comes from the same source.

We can now express the process by the following equation:

$$CO_2 + H_2O + \text{light energy}$$

$$\xrightarrow{\text{in presence of chlorophyll}} C_6H_{12}O_6 + O_2$$

But this simplified equation does not explain the various reactions which take place before the final products are formed. It is now known that there are two main stages. The first is the absorption of light energy by the chlorophyll, which changes it to chemical energy. The chemical energy is then used to split water molecules into hydrogen and oxygen.

$$2H_2O \rightarrow 4H + O_2$$

During the second stage, which can take place in the dark, this hydrogen reduces the carbon dioxide molecules to form a simple carbohydrate with the generalised formula (CH_2O):

$$CO_2 + 4H \rightarrow (CH_2O) + H_2O$$

This reaction is not a simple one and is catalysed by a series of enzymes. The final product is glucose sugar, $6(CH_2O)$, more correctly expressed as $C_6H_{12}O_6$. Hence photosynthesis, without including the intermediate reactions, can be represented by the following balanced equation:

$$6CO_2 + 12H_2O \xrightarrow[\text{light energy}]{\text{chlorophyll}} C_6H_{12}O_6 + 6H_2O + 6O_2$$

(As all the oxygen released is known to come from the water, it is necessary to include 12 molecules of water in the equation to account for the 6 molecules of oxygen.)

Finally, when the glucose is turned into starch, the reaction involves the removal of water. The reaction is catalysed by the enzyme starch phosphorylase.

$$nC_6H_{12}O_6 \xrightarrow{\text{starch phosphorylase}} (C_6H_{10}O_5)_n + nH_2O$$

where n varies according to the type of starch.

In what part of the leaf is sugar turned into starch?

> Take a very thin leaf which has been exposed to the light for some time and examine it in a drop of water under the high power of the microscope. A leaf from near the tip of the water plant *Elodea* is a good choice; the green chloroplasts which you will see round the edges of cells may be in a state of movement.
>
> Now lift the cover slip and add a drop of iodine before replacing. Any starch grains present will go blue-black. Where are they situated, all over the cell or in special places?

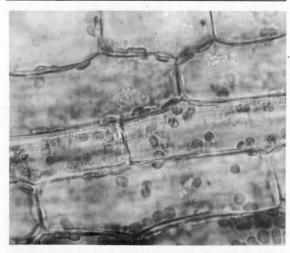

Fig. 15:8 High power photomicrograph of cells of an *Elodea* leaf showing chloroplasts.

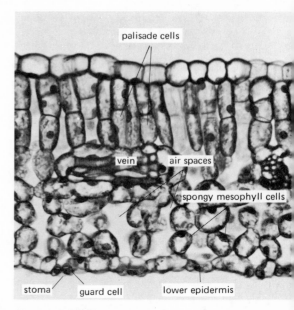

Fig. 15:9 Photomicrograph of a transverse section through the lamina of a leaf.

The leaf as an organ for manufacturing sugar and starch

Like all well-designed factories there must be good access for the raw materials, in this case carbon dioxide, water and light; good transport within it to bring the products together; and good means of removal of the final products, carbohydrates and oxygen.

Entry and transport of the raw materials

1. Stomata

Carbon dioxide enters the leaf through the stomata, which are usually situated on the underside. They are capable of opening and closing, and their movements are primarily affected by the intensity of the light and, to a lesser extent, by the evaporation of water. As a general rule it can be said that they open by day and close at night. The basic principle of the opening is that the two **guard cells** which surround each stoma act as osmotic systems and when the sugar concentration within them is high, water is drawn in and they enlarge. Because of special thickenings in their walls the guard cells can expand in certain directions only. As a result they move

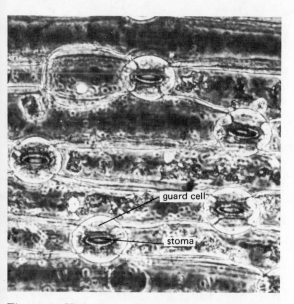

Fig. 15:10 High power photomicrograph of the lower surface of a leaf showing stomata.

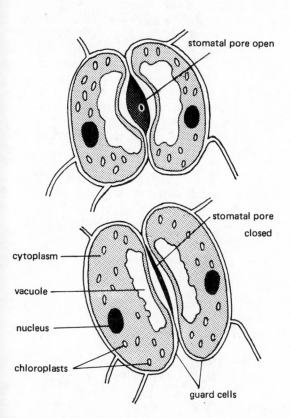

Fig. 15:11 Diagram of a stoma in surface view: (top) open, (bottom) closed.

apart making the pore between them larger (Fig. 15:11).
You can illustrate this action by making two model guard cells.

Take two pieces of Visking tubing about 10 cm long. Fill them with sugar solution, tying their ends together tightly as in Fig. 15:12. Each tube represents a guard cell. Place the tubes in a vessel containing water.

You now have an osmotic system as the walls of the tubing are semi-permeable. What happens to the space between the tubes after about an hour?

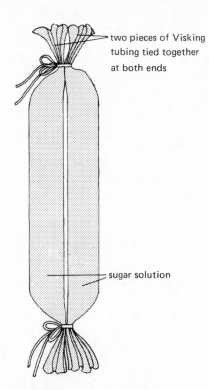

Fig. 15.12 Diagram of a model stoma.

To understand stomatal movement it is important to realise that the guard cells of a leaf are the only epidermal cells to have chloroplasts. Hence, when illuminated, they form sugar while their neighbours do not. Unlike the mesophyll cells, the sugar made in the guard cells is not turned into starch until night time. So during the day, the sugar solution in the guard cells becomes stronger, water is drawn in, and the stoma opens. At night,

153

when the sugar turns into starch, the osmotic strength of the sap in the guard cells becomes reduced and water is lost to neighbouring cells and the stoma closes. This is not the complete explanation regarding stomatal movement. There are still some factors involved which are not yet fully understood.

2. The passage of carbon dioxide from the atmosphere to the cells

Entry of carbon dioxide through the stomata takes place by diffusion. The process is very efficient because of the large surface area of the leaf and the vast number of stomata present. Once inside, the carbon dioxide passes to the mesophyll cells by diffusion within the intricate system of air spaces between the cells. It then dissolves in the water lining the air spaces, diffuses through the cell walls into the cytoplasm and so reaches the chloroplasts.

3. The transport of water

The water brought from the roots by the xylem is quickly spread to all parts of the leaf by a series of branching veins.

Hold up various leaves against the light and examine the network of veins with a hand lens.

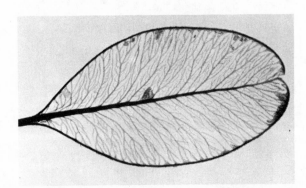

Fig. 15.13 A leaf 'skeleton' showing the arrangement of the veins.

4. How light reaches the chloroplasts

The efficiency of this process depends largely on two factors: the surface area of the leaf, and its position with respect to the light source. A large surface area allows more light to reach the chloroplasts and if the leaf is at right angles to the light, more will penetrate.

Consider some of the methods by which plants present as much leaf surface as possible. If a geranium is placed on a window ledge, how do its leaves react to light coming from the side? Next time you are in a wood, make a note of the ways in which the plants expose the maximum leaf surface to the light. Think in terms of the size and arrangement of the leaves and also notice to what extent they overlap each other.

Internally you will also find adaptations. Turn to Fig. 5:14 and see how the chloroplasts are arranged in the cells in relation to the light. Where are they most abundant? Remember that light gradually becomes absorbed as it penetrates the tissues, so a very thin leaf will be more efficient than a thick one.

Removal of the products of photosynthesis

The oxygen diffuses out through the system of air spaces and through the stomata. The sugar formed on the surface of the chloroplasts is turned into starch grains and accumulates in the leaf during the day. At night the reverse process takes place and the sugar in solution passes into the phloem for translocation to other parts of the plant. You will now understand why in some of the experiments involving the formation of starch the leaves were 'destarched' by keeping them in the dark first.

Thus the structure of leaves is admirably adapted to bring together the carbon dioxide, water and light as efficiently as possible and to remove the oxygen and the carbohydrates when they are produced.

What happens to the carbohydrates formed in photosynthesis?

Carbohydrates are used up in three main ways:

1. Respiration

Most cells receive sugar via the phloem; this can be broken down in respiration to liberate the energy the cells require.

2. Food storage

Some of the sugars may be stored for future use, either by the parent plant in special organs such as bulbs, tubers, corms or rhizomes, or for the next generation in seeds or fruits. The sugar is usually converted into starch for storage; occasionally it is stored as cane sugar, as in the roots of sugar beet. In some seeds it may be changed into an oil and stored as such. These storage organs are of great value to man because of their high concentration of food. Many of our vegetables are really plant food storage organs: potato tubers are storage stems; carrots, turnips, parsnips and beetroot are storage roots; onions are mainly composed of storage leaves; and beans and peas are seeds with large food stores. When oil is stored instead of starch this may be extracted for many purposes, e.g. olive oil from the fruit of the olive, and ground nut oil, among others, for the manufacture of margarine.

Examine the starch grains from various storage organs.

Take a scraping from the cut surface of a potato and mount in water with a little iodine. Draw the starch grains carefully. Repeat with scrapings from peas, rice and wheat mounted in the same way. What differences can you see?

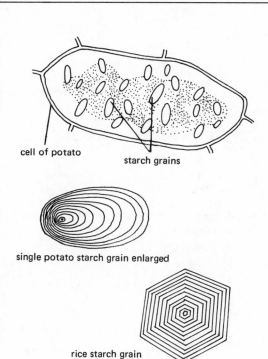

cell of potato   starch grains

single potato starch grain enlarged

rice starch grain

Fig. 15:14 Starch grains.

These grains are made in the cells of the storage organ from the sugar which reaches them from the leaves. The synthesis takes place within certain bodies which are like chloroplasts except that they contain no chlorophyll. These **leucoplasts**, as they are called, build up the starch grains, layer by layer, until they break through the outer membrane of the leucoplast into the vacuole of the cell. Sometimes the layers in the starch grain can be seen; did you see them in the starch grains from the potato?

3. Growth

Sugar is also taken to the parts of the plant which are actively growing. Growth results from an increase in the number of cells. Sugars are used in the formation of cellulose, a complex carbohydrate found in cell walls, fats and in the synthesis of protein, a basic component of protoplasm. All these processes are summarised in Fig. 15:15.

Protein synthesis

All proteins contain carbon, hydrogen, oxygen and nitrogen, and frequently sulphur and phosphorus as well (p. 161). Proteins are synthesised from smaller 'building blocks' known as **amino acids**. Plants, unlike animals, can synthesise amino acids by combining ammonia (formed from the reduction of **nitrates**) with certain complex chemicals derived from the breakdown of **sugars**. Sulphates and phosphates are also used in the synthesis of some proteins. All these chemical reactions are catalysed by enzymes (themselves proteins) and take place in the cell protoplasm.

Elements essential for healthy growth

Plants require a large number of elements if they are to grow properly. We can investigate the effect of these substances by depriving growing plants of some of them and seeing what happens. This can be done by using **water cultures**. This work was pioneered by the German scientists **Sachs** and **Knop** in the mid-19th century. They both devised culture solutions in which plants would grow healthily. Knop's solution is given in the second column of the table on p. 157.

The basic principle is that several chemicals are used in making up the water cultures. By

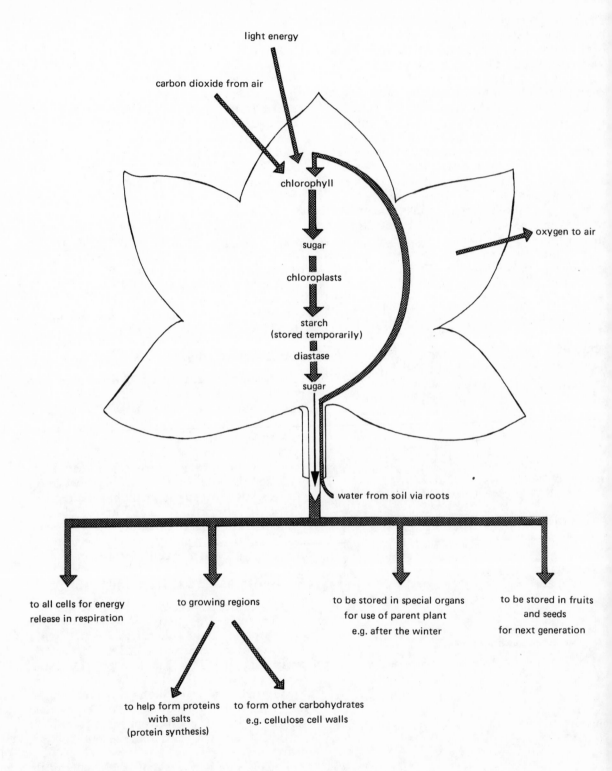

Fig. 15:15 Summary of the process of photosynthesis and the fate of the products.

using different combinations of these salts, solutions can be produced which are deficient in *one* element. For example, to produce a nitrogen-deficient solution, both calcium nitrate and potassium nitrate are removed. This takes away calcium and potassium as well as nitrate. However, potassium is still present as potassium dihydrogen phosphate, so only the calcium must be put back. This is done in the form of calcium sulphate (see table).

To see what happens when nitrogen, magnesium and iron are missing, proceed as follows:

Germinate some maize in moist sawdust, and when the seedlings are several inches high select five of *equal* height. Carefully wash off the sawdust and remove with a scalpel the whole of the remaining fruit (this contains food reserves, including minerals) without damaging the root or shoot. Transfer the seedlings to a series of gas jars which have been thoroughly cleaned and sterilised. Set up each jar as shown in Fig. 15:16 and label. One should contain the complete culture solution (the control), a second distilled water and the remainder the solutions deficient in nitrogen, magnesium and iron respectively. Surround the jars with black paper or aluminium foil to prevent the growth of algae in the culture solutions.

Keep the plants in the light. Bubble air gently into the solutions from time to time.

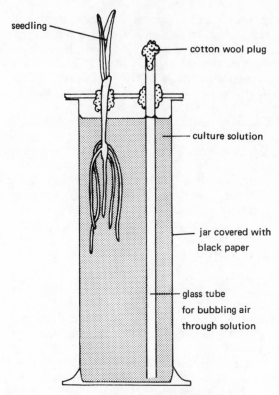

Fig. 15:16 Jar used in water-culture experiments.

(Why?) Observe the growth of the plants over a period of several weeks, recording their height, size, colour of the leaves, growth of the roots, etc.

Obviously, you cannot come to definite conclusions using only one plant in each cul-

TABLE OF CONSTITUENTS OF VARIOUS CULTURE SOLUTIONS

Substance	Complete Knop's solution	Lacking nitrogen	Lacking magnesium	Lacking iron
Calcium nitrate	0·8	—	0·8	0·8
Potassium nitrate	0·2	—	0·2	0·2
Potassium dihydrogen phosphate	0·2	0·4	0·2	0·2
Magnesium sulphate	0·2	0·2	—	0·2
Calcium sulphate	—	0·6	0·2	—
Ferric phosphate	trace	trace	trace	—

Note: All constituents are in grammes/litre. The solutions should be made up with distilled water.

ture solution. Far larger numbers are really necessary to overcome possible errors due to variation between the individual plants used. However, you should be able to establish in your experiment quite convincingly that the growth of plants is affected by the lack of certain elements.

Can you relate your results to the following facts?
1. Magnesium is part of the chlorophyll molecule.
2. Iron is used in the formation of chlorophyll.
3. Nitrate is essential for the formation of protein.
Why was one plant put in distilled water?

Trace elements

In addition to the main nutrients already mentioned, minute quantities of some elements are also needed by plants for certain metabolic processes. These are called trace elements and include boron, manganese, cobalt and zinc.

The role of the more important elements in plant nutrition can now be summarised:

Carbon, hydrogen and oxygen are obtained from carbon dioxide and water and are used in the formation of carbohydrates, fats and proteins. Nitrogen, phosphorus and sulphur

Fig. 15:17 Aerial view of the Broadbalk field, Rothamsted Experimental Research Station, Hertfordshire, showing experimental plots. The white area is a region that was left fallow.

are used in protein synthesis; iron and magnesium are necessary for the formation of chlorophyll; potassium is essential for cell division and calcium affects the permeability of cell membranes to solutions of salts. In land plants all these substances, except carbon dioxide, are taken up by the roots from the soil water.

How are minerals released from the soil?

When plants die their remains become incorporated in the soil as **humus**. This dead organic matter in the soil then decays due to the action of bacteria and fungi and as a result many essential nutrients such as nitrates are produced and returned to the soil.

Mineral salts may also be released from the underlying rocks by the action of soil acids and rain-water. The latter is also slightly acidic due to the carbon dioxide dissolved in it. The natural supply of mineral salts to the soil, however, is often greatly depleted by man through the growing of crops.

Natural and artificial fertilisers

When crops are regularly taken from the soil, the mineral salts locked up in the plants are not returned through the natural processes of death and decay. It follows, therefore, that crop yields will gradually diminish as more and more of these nutrients are removed, unless fertilisers are added.

In the famous Broadbalk Field at Rothamsted Experimental Research Station, wheat has been grown and harvested every year since 1843 without the addition of any fertiliser. Each year, for over 100 years, the yield has been measured, and during this time has decreased to about 45% of the original yield.

Farmyard manure is probably the best means of replenishing the soil but with the modern use of tractors, rather than horses, it is a scarce commodity. It takes time for the minerals to be released through bacterial action, but manure has the great advantage of preserving the soil structure through maintaining the level of humus. In a sandy soil the humus binds the sand particles together and helps to retain moisture; with a clay soil it tends to break up the fine particles, thus allowing water to drain away more easily.

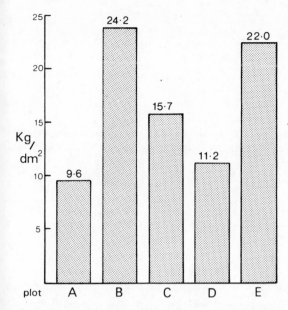

Fig. 15:18 Histogram of the average annual yield of wheat between 1852 and 1967 on five of the Broadbalk plots. A No treatment. B Farmyard manure. C Nitrogen only. D Potassium, phosphorus, sodium, magnesium (no nitrogen). E Potassium, phosphorus, sodium, magnesium, nitrogen.

Modern agriculture, with its intensive use of land, depends greatly on the use of artificial fertilisers. They are effective in promoting plant growth, but do nothing to maintain the soil structure. They are soluble and act quickly, but are easily washed out of the top soil by rain through a process called 'leaching'. Therefore they have to be added just before the seeds are sown.

Many of these chemical fertilisers are used, but the commonest include ammonium sulphate, ammonium nitrate, sodium nitrate, 'superphosphate' (mainly calcium phosphate), and potassium sulphate.

The choice of fertiliser depends largely on the crops being grown and the type of soil. Cereals, for example, require more nitrogen than other types of crops while potatoes need more potassium. Knowledge about this has been obtained only by painstaking research. The Broadbalk Field at Rothamsted was divided into eighteen experimental strips in 1843. Wheat was sown in each and treated in different ways. The yields were compared each year. A summary of the main results is given in Fig. 15:18. They give convincing evidence that certain combinations of fertilisers do increase yields very considerably.

Crop rotation

In order to economise on fertilisers and to preserve the structure of the soil, it is normal practice in many places for farmers to sow crops in a particular rotation over a period of three or four years; but it is also common, now that intensive methods are used much more, for two crops to be harvested in one year. However, the basic method of rotation is to grow cereals in the first year, a root crop such as beet in the second, and crops producing a lot of nitrogen, e.g. clover, peas, beans, in the third. Alternatively a four-year rotation might be: wheat—roots—barley or oats—clover, peas or beans. Fertiliser is added before the sowing of the root crop. Clover is usually sown with the barley or oats in the third year and harvested in the fourth; their roots are ploughed in before the next cereal crop is sown.

One advantage of a rotation is that pests, diseases and weeds tend to be prevented from establishing themselves; however, with the more recent development of pesticides and selective weedkillers, traditional patterns of growing crops are changing.

16

Food and diet

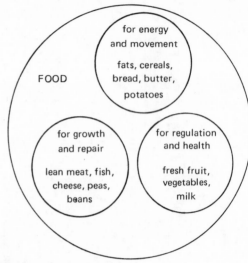

Fig. 16:1 Categories of food.

Why do we need food?

Food is a generalised term which includes all the materials taken in through the mouth which are necessary for the activity, growth and well-being of the body. We can divide these into the following categories:

1. Fuels to provide energy for movement and all the chemical processes going on in the body.
2. The raw materials for the growth of the body and the repair or replacement of tissues.
3. Substances concerned in regulating the chemical activities of the body.
4. Substances required for the health and protection of the body.

A good diet must not only include all these categories, but they should be in the right proportions for the person concerned. For man, these needs are met if the diet includes the following: carbohydrates, fats, proteins, mineral salts, vitamins and water. Our energy requirements are supplied by carbohydrates, fats and proteins; for growth and repair we need proteins, mineral salts and vitamins; for regulation and protection, mineral salts and vitamins are necessary, and water is needed for every process taking place in the body.

Carbohydrates

These provide energy when their molecules are broken down during respiration. They include starches and sugars and occur in large quantities in cereals, bread, potatoes and most root vegetables. They all have molecules composed of carbon, hydrogen and oxygen only, the hydrogen and oxygen being in the same proportion as in water, e.g. glucose sugar, $C_6H_{12}O_6$; sucrose, also called cane sugar, $C_{12}H_{22}O_{11}$; starch, $(C_6H_{10}O_5)_n$. The 'n' outside the bracket may be a number between 300 and 1000, the figure depending on the type of starch. In other words a starch molecule is like a string of beads, each bead being a $C_6H_{10}O_5$ unit. When starch is turned into sugar by digestion, this molecule is broken down into its units and a water molecule is added to each. The reaction takes place in two stages, but can be summarised as follows:

$$(C_6H_{10}O_5)_n + nH_2O \xrightarrow{\text{enzymes}} nC_6H_{12}O_6$$

Glycogen, a substance rather like starch, is an important fuel reserve in man; most of it is stored in the liver.

Fats

These include oils, a term used for fats which have a low melting point and hence are liquid at room temperatures.

Fats are also energy foods. Their molecules, like those of carbohydrates, consist of carbon, hydrogen and oxygen atoms only, but they differ from them in *not* having the hydrogen and oxygen atoms in the same proportion as those in water.

160

Fats are present in almost pure form in butter, margarine, olive oil and cooking fat, while meat, milk, egg yolk and nuts contain appreciable amounts.

Fat is easily stored in the body under the skin and around such organs as the kidneys. When stored under the skin, it acts as an insulating layer which helps to keep in the heat of the body: the blubber of whales is an extreme example of this. Stored fat also serves as a food reserve. As such, it is particularly important for hibernating mammals such as hedgehogs and dormice as they are dependent on fat to keep them alive during the winter when food is scarce.

Fat is also important to us as it acts as a vehicle for the intake of vitamins; many vitamins are only soluble in fat.

Proteins

These can also be used as a source of energy—indeed carnivores such as lions rely largely on them for this—but their unique and essential role is for the growth and repair of the body tissues. Without protein, nothing grows.

Proteins are extraordinarily complex chemical substances, containing atoms of carbon, hydrogen, oxygen and nitrogen and often sulphur and phosphorus as well. These huge molecules are built up from smaller units called amino acids which may be combined together in various ways; theoretically there is an almost infinite number of possible proteins that could be formed from them. Every animal species synthesises its own particular kinds of proteins from the amino acids it obtains from its food. Our own body requires about twenty amino acids; at least eight of these must be obtained from proteins in the diet—the remainder can be synthesised. Proteins derived from animal sources such as lean meat, fish, cheese, eggs and milk are extremely good sources of these eight essential amino acids. Some plant proteins derived from peas, beans and oil seeds compare favourably with them, but plant proteins on the whole are more likely to provide only some of the essential amino acids.

The daily needs of an adult are not less than 70 g of protein, of which about half should be derived from animal foods. Those who prefer a more vegetarian diet can take more of such protein-rich foods as milk, cheese and yoghurt. More protein is needed during periods of quick growth, which means that a teenager usually requires more than his father; more is also needed during pregnancy and breast-feeding. Elderly people, especially when living alone, often tend to cut down on proteins; although they need less, proteins are nevertheless vital for maintaining their tissues in a healthy state.

Testing food substances

You can find out whether carbohydrates, proteins and fats are present in foods by carrying out a series of tests. It is interesting to analyse such foods as sausage, onion, potato, bread, soaked dried peas and cheese. Proceed as follows:

1. Grind up the particular food to be analysed with a pestle and mortar. Put a little of the product into a clean, dry test tube, add 2 cm^3 of ethanol and shake. If fat is present some of it will dissolve in the ethanol. Allow the remaining contents of the tube to settle, decant off some of the clear liquid into another test tube and add about 2 cm^3 of cold water. It will go milky if *fat* is present; fat is insoluble in water and so it comes out of solution as fine droplets, known as an **emulsion**.
2. Take the remainder of the ground-up food and mix it with water; some substances will dissolve, the rest will be suspended. Use this mixture as stock material, taking about 2 cm^3 of it for each of the following tests:
a) Add a few drops of iodine dissolved in a solution of potassium iodide. A blue-black colour indicates that *starch* is present.
b) Boil the sample for one minute with 2 cm^3 of Benedict's solution in a boiling tube; move the tube in and out of the flame carefully so the liquid does not spurt. A colour change from blue to green to orange and finally the formation of a brick red precipitate, indicates the presence of a *reducing sugar*, e.g. glucose. If only a slight change of colour occurs, reducing sugar *is* present, but only in very small amounts.
c) If *no* reducing sugar is present, boil the sample for two minutes with three drops of dilute hydrochloric acid in a boiling tube, cool the tube under a tap, then add *small*

quantities of solid sodium bicarbonate until the fizzing stops (this indicates that the acid is neutralised). Add 2 cm³ of Benedict's solution and boil. A colour change as in b) indicates the presence of a *non-reducing sugar*, e.g. sucrose.

d) Add a few drops of Millon's reagent and heat gently; a pink or red colour indicates the presence of *protein*. This reagent is poisonous so it should be handled with care.

Compile a table showing which substances are present in the various food items analysed by the whole class.

Minerals

Mineral salts are essential because they provide many of the elements needed for growth, protection and the regulation of metabolic processes. Of these, elements such as sodium, potassium, calcium, phosphorus and iron are required in appreciable quantities.

Sodium, in the form of sodium chloride, is necessary to maintain the right osmotic balance of body fluids such as blood, hence enough has to be taken in each day to replace what is lost in sweat and urine.

Calcium is needed for muscle contraction and the formation of strong bones and teeth (along with phosphorus, magnesium and vitamin D). It is especially important during pregnancy as more is then required for the growth of the foetus. If insufficient calcium is present in the mother's blood at this time, the blood will extract some from the mother's bones and teeth which will suffer in consequence. A high calcium intake is also necessary when breast-feeding because calcium is an essential ingredient of milk.

Phosphorus is needed for bone and teeth formation, but in addition it is necessary for the adequate absorption of calcium from the gut. It is also an essential ingredient of ATP (p. 130), the substance used for transferring energy during the respiratory process.

Iron is essential for making **haemoglobin**, the red, oxygen-carrying pigment in the blood corpuscles. If iron supplies are inadequate **anaemia** results. Anaemia is a condition in which the haemoglobin content of the blood is below the normal for good health; the person becomes pale and lethargic and may have periods of breathlessness and dizziness.

Iodine is only needed in very small amounts, but it is essential because it is the active component of thyroxin, a growth-controlling substance formed in the thyroid gland.

Minerals

Element	Good sources of supply	Importance
Calcium	Milk and cheese especially, also green vegetables and eggs.	For the development of strong bones and teeth. For muscle contraction. For clotting blood.
Phosphorus	Meat, fish, eggs. Wholemeal bread.	For the development of strong bones and teeth. For the absorption of calcium from the gut. For making ATP.
Iron	Liver, meat, eggs. Wholemeal bread. Water-cress, spinach and other green vegetables.	For making the haemoglobin in the red corpuscles of the blood.
Iodine	Mainly in sea-water and seaweed, but also present in various sea foods. Often found in table salt.	For making thyroxin which is essential for growth.
Fluorine	Drinking water in some regions only.	Hardens the enamel of teeth and helps to prevent dental decay.
Sodium	Most foods.	For maintaining the osmotic balance of body fluids.
Potassium	Most foods.	For normal cell function and nerve action.

In addition there are certain essential **trace elements**, so called because they are only needed in extremely small amounts; these include fluorine, copper, manganese and cobalt. Their importance can be illustrated by fluorine. In some places fluorine is present naturally in drinking water, but in others it is absent or insufficient. It has now been established by carefully controlled experiments that by adding fluorine to the water supply at a concentration of two parts per million, dental decay, especially in school children, is very greatly reduced. This is because the fluorine has a hardening effect on the enamel of the teeth. In Slough, for example, when the water supply was treated in this way the percentage of children with perfect teeth rose from 5 to 30%, and in South Shields dental decay was reduced by 44% compared with North Shields where no fluorine had been added to the water.

The main sources from which we obtain these minerals are given on the previous page.

Vitamins

If rats are fed on a diet of pure carbohydrate, protein, fat, mineral salts and water they will quickly die, but if a few cm^3 of milk are added daily they will thrive. This is because milk contains **vitamins**. Vitamins are organic substances which are found in a great variety of foods and are essential for life and health; they are required only in extremely small amounts.

The importance of vitamins for the health and protection of the body was realised long before their chemical composition was known, so for convenience they were called by letters of the alphabet. Today they are often referred to by their chemical names, e.g. vitamin C is ascorbic acid. Vitamins are important substances because many of them are involved with enzymes in vital metabolic processes.

Most vitamins are synthesised by plants (including bacteria), although mammals can partially synthesise vitamins A and D from substances of plant origin. We obtain vitamins either directly by eating plants, or indirectly from animals which have acquired them from plants.

Diseases caused by lack of vitamins are called **deficiency diseases** to distinguish them from diseases caused by viruses or bacteria. Scurvy, beri-beri, pellagra and rickets are examples. Scurvy was a common disease among sailors in the days of long sea voyages, but **Sir Richard Hawkins**, as long ago as 1593, realised that the best treatment for scurvy was eating oranges or lemons, although it was not until the twentieth century that these were known to be effective because they contained vitamin C.

The table on p. 164 summarises the more important facts about the various vitamins. Examine it carefully and note particularly the best sources of supply, because these should be present in a good diet. It is now a usual practice to enrich certain foods by adding vitamins; you can read about this on such things as breakfast cereal packets.

Gross vitamin deficiency is becoming unusual in countries where there is a high standard of living and the people have good mixed diets, but it has by no means been completely eradicated. Deficiency diseases are still quite common amongst the very poor and especially in aged people who live alone. In many parts of the world, vitamin deficiency is a very serious form of malnutrition (p. 167).

Testing for vitamins

Most of the vitamins are very difficult to test for, but vitamin C (ascorbic acid) decolorises a blue dye called dichlorophenol-indophenol, DCPIP for short. This is because ascorbic acid is a powerful reducing agent. Try this test for yourself:

Place 1 cm^3 of freshly prepared DCPIP solution in a test tube. Now take up 1 cm^3 of a 0·1% solution of pure ascorbic acid (made up from a tablet) in a syringe and add it drop by drop without shaking the solution until it is decolorised.

You can now use this test to see whether ascorbic acid is present in a variety of foods, provided they are not strongly coloured. Include both raw and cooked fruits and vegetables. Grind up each type of food with a pestle and mortar and shake with water. Any ascorbic acid should go into solution. Add this extract (filtered if necessary) to 1 cm^3 of DCPIP as before and note whether it becomes decolorised or not. Summarise the results from the class by making a table showing

163

which foods contain ascorbic acid. Did cooking the food make any difference?

Water

Water is an essential part of our diet for many reasons: digestion and absorption of food could not occur without it; it is necessary for transporting material within the body; all metabolic processes take place in a watery medium and water is an essential ingredient of protoplasm itself.

Our bodies consist of up to 76% water by weight and, on average, a person loses about 3000 cm^3 of water every day as urine and through evaporation from the skin and lungs. This water must be replaced by what we drink and by the water in our food.

Vitamin Chart

Vitamin	Principal food sources	Functions	Deficiency effects	Other points
Fat soluble				
A	Milk, butter, eggs, fish-liver oils; green vegetables, tomatoes, carrots. (These plants provide carotene from which vitamin A is synthesised.)	Keeps surface and lining tissues healthy. Essential for normal growth.	Infection of eyes, nose, throat and skin. Night blindness (inability to see in dim light). Slows down growth.	Can be stored in the body. Not destroyed by cooking.
D	Fish-liver oils, milk, butter, eggs. (Also formed in the skin by the action of sunlight.)	Concerned with the uptake of calcium and phosphorus.	Rickets, a disease found mainly in children, where bones fail to harden properly and consequently bend.	The vitamin is found in several forms. Rickets occurs mostly in children who receive little sunshine, e.g. in large cities.
E	Wholemeal wheat, eggs, butter.	Importance to man uncertain.	Sterility in some mammals.	Seldom deficient.
Water soluble				
Vitamin B complex:				
Thiamine (B$_1$)	Yeast, wholemeal bread, liver, peas and beans.	Essential for growth and general health.	Beri-beri, a nervous disease.	Beri-beri common in countries where polished rice is the main cereal in the diet.
Riboflavine (B$_2$)	Yeast, wholemeal bread, meat, milk, eggs, peas, green vegetables.	Keeps skin healthy.	Dermatitis (skin disease).	
Nicotinic acid	Yeast, liver, eggs, milk, cereals.	Concerned with health of skin, digestive and nervous systems.	Pellagra (a disease affecting the skin, alimentary canal and nervous system).	Pellagra common where maize is the main cereal in the diet; can be reduced through adding wheat flour.
B12	Liver.	Not fully understood.	Pernicious anaemia (deficiency of red blood corpuscles).	
C (Ascorbic acid)	Blackcurrants, oranges, lemons, grapefruit, tomatoes, fresh green vegetables.	Essential for health of skin.	Scurvy (bleeding from gums and under skin, weakness, poor healing quality of bones).	Easily destroyed by cooking.

Roughage

This is the term used for the indigestible material taken in with the food; it consists largely of cellulose and lignin from plant foods. Although not strictly a nutrient, roughage is a necessary ingredient of food, because by adding bulk it stimulates the muscles of the gut wall to contract more vigorously, thus keeping the food on the move and preventing **constipation**. Constipation is the retention of the faeces within the gut for longer than normal, a condition which, if habitual, may lead to ill-health. Much 'processed' or 'purified' food lacks roughage, but fruits, salads, vegetables and wholemeal bread contain a good supply.

Fuel requirements

The amount of energy released from food can be estimated using the apparatus shown in Fig. 16:2. A known weight of the food is burnt in a current of oxygen and the heat given off is absorbed by a known volume of water. From the rise in temperature of the water, the energy released from the food can be calculated. Previously this energy value was expressed in kilocalories (1 kcal being the amount of energy required to raise 1 kg of water through 1°C); it is now expressed in kilogramme joules (kJ). The conversion is 1 kcal = 4·19 kJ.

It has been calculated that a man needs on average to eat enough food to provide him with 7123 kJ of energy just to keep him alive. This is used for all the metabolic processes going on in the body and for keeping his temperature constant. The amount required over and above this figure varies considerably according to his activity (and hence his occupation) and his age. On average, women require about 10% less than men, but more is needed during pregnancy and when breast feeding a baby. Fig. 16:3 gives some approximate comparisons.

Why do you think more energy is needed during the period of adolescence (13–15)? What difference do you think climate would make to these figures? Consider the surface area of the body in terms of heat loss (and therefore more energy needed to replace it); how would the energy needs of men of the same age differ according to height and build?

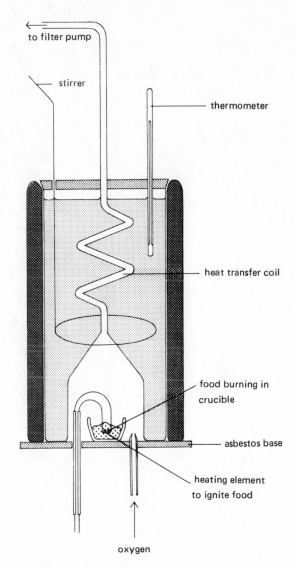

Fig. 16:2 Food calorimeter.

All the energy a person needs is provided by the carbohydrates, fats and proteins that are eaten; if insufficient food is taken in, a person will lose weight because his reserve foods will be used up and even his tissues will be used as fuel for oxidation. If he eats too much he may put on weight. In Britain, in 1970, the average person ate 9% more energy food than was necessary! People in different countries tend to obtain the energy they require from different foods according to availability and climate. Until recently, Eskimos, for example, ate far less carbohydrate

165

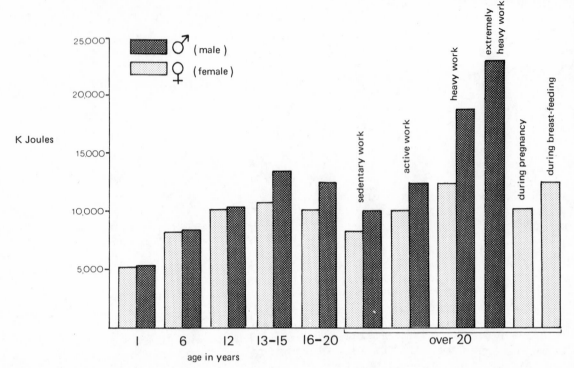

Fig. 16:3 Human energy requirements.

and vastly more fat and protein than people in other parts of the world. Why was this?

Finding out the energy value of a meal

Find out for yourself the energy value of a simple meal, e.g. fish and chips, or a picnic lunch of ham or cheese sandwiches.

To do this you have to calculate the total weights of carbohydrates, fats and proteins from all the items. First weigh each item on the menu separately, e.g. the bread, butter and cheese from the sandwiches. Then calculate the weight of carbohydrate, fat and protein in each item by referring to the table opposite. Finally, calculate the total weight of each class of food in the full meal and work out the energy values on the basis that:

1 g of carbohydrate releases 17·2 kJ of heat energy; 1 g of fat releases 38·5 kJ and 1 g of protein, 22·2 kJ.

Summary of the essential requirements of a good diet

1. It should contain enough energy food to meet the requirements of the person concerned.
2. It should contain enough protein to provide the right amino acids for the growth and repair of the body tissues. (These two requirements are met daily by eating about 70 g of proteins, 100 g of fat and 500 g of carbohydrate.)
3. It should be varied to provide adequate salts and vitamins.
4. It should contain enough roughage to stimulate vigorous peristalsis.

A day's menu

Work out a good balanced menu for yourself to suit your own requirements. Refer to the table opposite and keep in mind the foods which provide the minerals (p. 162) and vitamins (p. 164). As a guide, here are some of the best foods arranged in categories; a good diet

PERCENTAGE COMPOSITION BY WEIGHT AND THE ENERGY CONTENT OF VARIOUS FOODS

Food		Protein	Percentage composition of Carbohydrate	Fat	Water	Energy (kJ/g or cm³)
MEAT AND POULTRY	Beef	23·5	—	20·4	54·8	12·0
	Roast mutton	25·0	—	22·6	50·9	13·1
	Bacon	9·9	—	67·4	18·8	27·9
	Pork	22·5	—	21·0	49·2	12·0
	Chicken	21·5	—	2·5	74·8	4·6
FISH	Cod	16·5	—	0·4	82·6	2·9
	Herring	19·5	—	7·1	72·5	6·1
	Salmon	22·0	—	12·8	64·6	8·8
	Trout	19·2	—	2·1	77·8	4·1
FATS	Butter	1·0	—	85·0	11·0	33·2
	Lard	—	—	100·0	—	38·8
DAIRY PRODUCE	Cheese	29·6	—	38·3	28·6	20·0
	Milk	3·3	5·0	4·0	87·0	3·0
	Eggs (boiled)	13·2	—	12·0	73·2	7·1
FRUIT	Apples	0·4	14·2	0·5	84·6	2·7
	Bananas	1·3	22·0	0·6	75·3	4·2
	Oranges	0·8	11·6	0·2	86·9	2·2
	Grapefruit	0·6	5·7	0·1	93·6	1·1
VEGETABLES	Potatoes	2·5	20·9	0·1	75·5	4·0
	Beans (dried)	22·5	59·6	1·8	12·6	14·8
	Peas (dried)	25·0	60·0	1·0	11·0	14·0
	Cabbage	1·6	5·6	0·3	91·5	1·3
	Lettuce	1·2	2·9	0·3	94·7	0·9
MISCELLANEOUS	Flour (white)	11·4	75·6	1·0	11·5	15·3
	Bread (white)	9·3	52·7	1·2	35·6	11·1
	Biscuits	9·0	70·0	10·0	11·0	18·0
	Honey	4·0	81·0	—	15·0	13·0
	Sugar (white)	—	100·0	—	—	17·1
	Chocolate	12·9	30·3	48·7	5·9	26·4

would contain daily something from each group.

1. Meat, fish, cheese, egg, beans, peas.
2. Butter, margarine.
3. Milk.
4 Potatoes, root vegetables, onions.
5. Cereals, bread.
6. Green vegetables (cooked or raw).
7. Tomatoes, oranges, grape fruit.

When considering diet it is also important to remember that even if you choose what to eat carefully, according to the correct principles, the beneficial effects of the diet may be influenced by other circumstances. If food is attractively prepared and eaten in a relaxed atmosphere it will be digested better. If snacks of poor nutritional value are taken between meals the appetite may be reduced for the more important meals.

Diet in different parts of the world

More than half the world is very poor by Western standards, so the food eaten is the cheapest available to satisfy hunger. The cheapest food, which is the food most easily produced, is vegetable in origin and consists largely of carbohydrates, chiefly starches. Cereals and root crops come into this category, the type varying according to the climate and to the traditions of the people.

In many parts of the world cereals and root crops are just about all that are eaten with the result that there is a gross deficiency of protein (especially of animal origin), vitamins and mineral salts; in other words it is not a *balanced* diet. When any of these three vital constituents are lacking a state of **malnutrition** exists. It is estimated that over half the world population suffers in this way. About 10% do not

Fig. 16:4 Map showing regions where kwashiorkor occurs (shaded areas).

Fig. 16:5 Child suffering from kwashiorkor feeding on high protein diet in hospital.

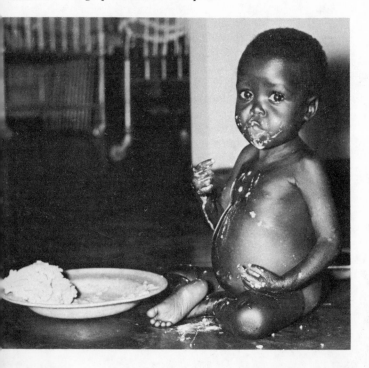

even obtain enough energy for their minimum needs and are actually wasting away; this is **starvation**, the cause of death of vast numbers of people every year in countries like India. Millions of others exist on a diet producing less than 8,400 kJ.

Protein deficiency is common in India, Tropical Africa, the Nile basin, and parts of Central and South America, to name some of the more important regions. These are often areas of very dense population and deficiency may be due to a) dependence on crops which contain very little protein such as cassava and bananas, b) low production of domestic animals or fish, c) poverty. Protein is always expensive to buy, especially animal protein.

Protein deficiency in children is the cause of **kwashiorkor**, a horrible disease associated with stunted development, wasting of the muscles and gross swelling of the tissues due to their water-logged condition. The disease often develops in the child's second year, because when breast-fed he receives enough protein, but when another baby is born he is weaned to a diet consisting mainly of carbohydrate, and the symptoms begin to appear. In some parts of Tropical Africa up to 10% of the children suffer in this way.

This should be contrasted with the situation in many of the richer nations where over-eating causes more diseases than under-eating. As income increases, so the intake of animal protein becomes larger and excess energy food is laid down as fat. However, obesity (the state of being fat) is not always entirely due to excess feeding, as some people accumulate fat much more readily than others, but being sensible about diet is important because fatness brings great disadvantages. Fat people are less healthy and have a shorter life expectancy; the more weight there is to carry about, the more strain on the heart and the greater likelihood of high blood pressure and heart attacks. Fatness limits activity and good exercise is a basic need for health. Nobody likes being fat and very few need be. Twenty per cent of people in the United Kingdom are overweight, partly because they eat too much and partly because they eat the wrong kind of foods. Many older people become fat largely because they continue to eat the same quantity as when they were much more active. However, where there is sufficient determination and self-control, it is possible to prevent, or at least reduce, the condition. To slim, it is necessary to eat smaller amounts of high energy foods, especially those containing starches and sugars, but it is essential to maintain a properly *balanced* diet. You could work out the effect of cutting out snacks such as sweets, biscuits, cakes, ice cream and potato crisps!

Intake of sugar alone is a very important factor in the diet of modern man. In the

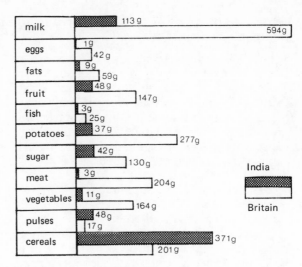

Fig. 16:7 Comparison of average Indian and British diets.

United Kingdom the average consumption of cane sugar has gone up from 6·75 kg to 54 kg per head per year since the early part of the last century. This has not only contributed to obesity and its attendant ills, but has had various physiological effects which have led to an increase in such conditions as diabetes, peptic ulcers and dental decay. The following is an example of the type of evidence which suggests this association of disease with excess cane sugar:

a) Sugar consumption among Natal Indians is nine times higher than among Indians living in India, and the incidence of diabetes is more than ten times as high in the former.

b) In rural districts of Natal it has been estimated that Zulus consume 2·7 kg of sugar a year compared with 38·25 kg in cities such as Durban where Zulus eat a more Western-style diet. In a large hospital which serves the rural districts there were only ten cases of diabetes out of nearly 10,000 admissions, while a Durban clinic recorded 1600 Zulu diabetics.

One important line of research on diseases is to map out their incidence on a world map; this often shows that they are characteristic of certain regions. The big question in each case is why that should be. Is it due to climate, to the presence of certain insect carriers, to the diet of the people, or to any other cause? It seems likely that diet plays a large part in the occurrence of some diseases.

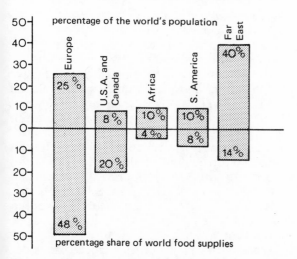

Fig. 16:6 How the world's food is shared.

169

What can be done about the world food situation?

The Food and Agriculture Organisation (FAO) of the United Nations was set up in 1962 to take practical steps to meet the dietary needs of people in various parts of the world, and in addition many voluntary organisations have made very valuable contributions towards solving the problem. The following are some of the lines being pursued:

1. Distributing food from places of plenty to places of need

It seems obvious that if there is a surplus in one place it should be sent to regions where it is needed. This has happened with surplus wheat in America and has been organised during times of famine to good effect, but it only provides temporary assistance.

2. Increasing the amount of land used for cultivation

Only about 10% of the land surface of the world is under cultivation. Obviously large areas are impossible to cultivate, but much more could be used. Even deserts can be cultivated if irrigation and sand consolidation methods can be applied. The high cost is the chief problem here.

3. Using better farming methods

Many methods are similar to those used for centuries. Yields are poor for the amount of labour exerted. Better methods are available if people can be persuaded to use them, and aid to developing countries in the form of machinery and fertilisers has already revolutionised yield in some parts. But advanced machinery is not necessarily the answer as this tends to produce unemployment; it is often better to use human labour to better effect. Mixed farms are more productive than those that specialise in one product such as cereals, as in the former more animal food is produced and more manure becomes available to put back on the land for the growing of crops.

4. Use of new varieties of cereals, sheep and cattle

By crossing different strains and selecting useful varieties, new types of wheat, rice and maize have been produced which have brought about a 'green revolution' in the past decade. For example, in Pakistan between 1966–71 the yield of wheat increased from 4 to 8 million tonnes per annum and in India during the same five years it rose from 12 to 20 million tonnes. In a similar way varieties of sheep and cattle have been developed which, when given protein-rich diets, mature much earlier; hence more food for human consumption is produced in a given time. Research is going on continually to raise better varieties of crops, fruit and farm animals, suitable for growing or rearing under different conditions in various parts of the world.

5. Increasing supplies of protein
a) By increasing the fish supply

It has been estimated that the world fish harvest could be increased by as much as 50%. Off some shores much more fishing could take place, but in the waters near highly developed countries over-fishing is already occurring and conservation measures will have to be taken to ensure a continued supply.

Fish-farming is carried out successfully in some countries and could be further developed. An interesting experiment was carried out in Scotland whereby nutrients were added to the water of a sea-water loch. This caused the growth rate of plaice and flounders to be significantly increased. This method could be successful in other localities where the water is more or less confined to a limited area.

Fig. 16:8 Fish farming in the Philippines.

Fig. 16:9 Irrigating a sugar plantation, E. Africa.

Fig. 16:10 Combine-harvesters at work, Canada.

b) By using indigenous animals

In some countries, notably in Africa, native animals can be used as a more profitable source of protein than the imported cattle. These animals, in contrast to cattle, have been in the country for millions of years and have become well adapted to conditions there. They are largely immune to the local diseases, can extract more nourishment from the vegetation and need far less water. By a regular 'cropping' technique numbers can be kept at an optimum for the food available. One species of antelope, the eland, has been successfully domesticated and used both for milk and meat.

c) By producing artificial meat

Synthetic meat is produced from soya beans and is already on the market; it is also being developed from a fungus. It looks and tastes like meat, costs only a fifth of the price, is easily transported and has a high nutritional value.

6. By enriching food

Many basic foods have nutrients added to them to prevent malnutrition. Vitamins and calcium salts are added to bread and other vitamins to margarine. In developing countries dried skimmed milk is the commonest supplement. A biscuit made from wheat flour and ground nuts has been used on a large scale in India, Singapore and Fiji; its high protein content has greatly benefited the health of children in these places.

7. By protecting food

When food is stored it is often attacked by beetles, rodents, moulds and bacteria. In parts of India, for example, this may reduce the food for human consumption by as much as 50%. Better methods of pest control, storage and transportation could dramatically increase world supplies.

171

17

Feeding methods

Place *either* a live sea mussel in a small dish of sea water, *or* a freshwater mussel in fresh water. Leave it undisturbed until the two valves of the shell open, and the two siphons concerned with water flow are seen projecting between them (Fig. 17:1). Sprinkle some mud particles in the water and observe how they are drawn towards one of the siphons.

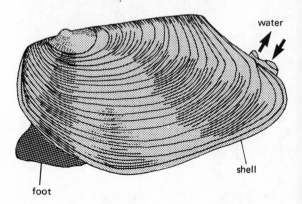

Fig. 17:1 Fresh-water mussel: a filter feeder.

Before food can be utilised by any animal it has to be obtained from some source outside the body and ingested, in most animals, through the mouth.

Methods of obtaining food

When we eat our food we use different tools for different substances. We would find knives unsuitable for peas and chopsticks for soup! In the same way animals feed on a great variety of food and many methods of capture and ingestion have been evolved to cope with particular problems.

Invertebrates obtain their food in many different ways. You will recall how an amoeba engulfs food particles by enclosing them by means of pseudopodia (p. 13), how a tapeworm absorbs digested food through its surface by diffusion (p. 37) and how the mouth parts of insects are adapted for biting, licking, sucking, or piercing and sucking (p. 69).

Another method of obtaining food, which is used by a great variety of animals, is known as **filter-feeding**. This method is typical of many animals which feed on very small organisms, or particles of organic matter in the water. Filter feeders are characteristically **sedentary** animals, i.e. they move very little or not at all; sponges, and bivalves such as mussels, are examples. The latter draw a current of water through their bodies by means of vast numbers of cilia and the food particles in the water are sieved off and used as food. You can observe how a bivalve mollusc does this by examining one in water.

A single sea mussel draws at least 11·25 dm³ (2·5 gallons) of water through its body every day. Many mussel beds contain tens, sometimes hundreds, of thousands of individuals; so you can calculate the vast quantities of sea water that must be filtered by such mussel beds.

Although most filter-feeders are invertebrates, the true whales also feed in this manner. These huge mammals feed largely on small shrimp-like crustaceans which occur in the plankton. When a whale feeds, it opens its jaws and takes in a great quantity of water containing planktonic organisms. It then nearly closes its jaws and, with the help of its tongue, forces the water out through a sieve of **baleen** or whalebone, which hangs from the edge of the upper jaw all round the mouth. It then swallows the plankton which has been sieved off (Fig. 17:2).

Other vertebrates use many specialised methods of obtaining their food. We have seen how birds use beaks of many shapes and sizes according to the types of food eaten and the manner of obtaining it (p. 102). Mammals also vary greatly in their techniques. Consider first those which graze. If you observe sheep and cows grazing you will notice that a sheep

172

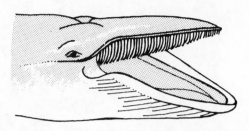

Fig. 17:2 Head of a fin whale showing fringe of baleen for sieving plankton.

uses its teeth to *bite* off the grass while a cow uses its tongue to *pull* it up. How would this difference determine the length of grass on which the two species could feed? What implications would this have for the farmer? The elephant also grazes, but it uses its trunk, which is an elongated nose, to grasp the vegetation and transfer it to the mouth. An elephant also uses its trunk when drinking: first it sucks water into its trunk, then it bends the trunk round until the end reaches its mouth so that it can squirt the water down its throat.

The carnivorous mammals use different techniques. Pack animals such as wild dogs and wolves run down their prey, using their teeth to bite them to death. They do not chew their food, but bolt huge pieces as quickly as possible before another member of the pack can rob them. Members of the cat family, which includes lions and leopards, leap on their prey and kill them with the help of their canine teeth; their rough tongues also enable them to rasp off the meat from the bones.

The anteater feeds almost exclusively on ants and termites and has to eat great numbers to supply its needs. Its long sticky tongue is an

effective organ for catching them. You could think out many more examples of the specialised ways by which other vertebrates obtain their food.

We have already mentioned the importance of teeth in capturing prey; they are also used for preparing the food before it is swallowed. We will now study teeth in more detail.

TEETH

How teeth have evolved

The teeth of all vertebrates are basically similar in structure to spines, called **dermal denticles**, found on the skin of some fish, e.g. dogfish, skates and rays.

> Examine the dermal denticles on a dogfish. Stroke the back of the fish from the tail end forwards. The denticles give the skin a roughness resembling coarse sandpaper. Examine the mouth region. What happens to the denticles as the skin passes over the jaws?

In amphibians and reptiles the teeth have remained all alike, as in these animals they are only used for gripping the food. However, some snakes have certain teeth which are enlarged to form fangs. The snake uses these fangs to bite its prey and pass poison into it (Fig. 17:4).

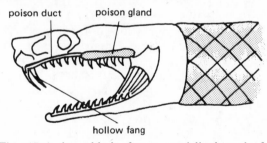

Fig. 17:4 An adder's fangs—specialised teeth for poisoning prey.

Fig. 17:5 Section through dogfish skin showing dermal denticles.

Fig. 17:3 An elephant uses its trunk to grasp vegetation and put it into its mouth.

Birds have no teeth as their function has been taken over by beaks, but mammals have teeth of various shapes and sizes which serve many different purposes. These range from the huge tusks of elephants, the largest known weighing over 100 kg, to the minute needle-sharp teeth of the shrews.

We will start our study of teeth by studying our own.

Human dentition

The term **dentition** is used to describe a complete set of teeth. We can find out something about our dentition by feeling our own teeth and looking at our neighbour's! We can also make a cast, as dentists do when they are fitting false teeth.

Take a piece of dental plaster big enough for all the teeth on one side of the mouth to bite upon. Now gently bite on it (do not bite right through it). When the plaster is nearly set, peel it off carefully. You should now have the impressions of your upper and lower teeth on the two sides of the cast.

Find out how many teeth you have on the upper and lower jaws, making sure that you are counting the impressions on *one* side of the mouth only. Do the upper and lower teeth meet exactly or not? What does this tell you about their functions?

Examine the impressions of the upper teeth first. You should have two **incisors** in the front. What is the function of their flat chisel-shaped edges? Next to the incisors is a single **canine**. Does it project more than the incisors? Would this tooth help if you were trying to tear something tough? Behind the canines are the cheek teeth which are of two kinds, the two **premolars** in front and the three **molars** behind. The last molar is called a wisdom tooth; you will probably not have any yet, as they come much later than the others. Premolars and molars help to crush and grind the food. Put your finger between your molars and bite gently; note the great power they can exert. This power increases the nearer these teeth are to the angle of the jaw; nutcrackers work on the same principle. Compare the contact surfaces of the molars with those of the canines and incisors. How does this make them more effective?

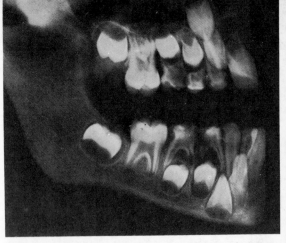

Fig. 17:6 X-ray photograph through the jaws of a 7-year old child showing permanent teeth embedded in the jaw below the milk teeth.

The difference between premolars and molars, apart from their position, is that during life there is only one set of molars, but there are two sets of premolars. The first set of teeth is called the **milk dentition**. At birth these teeth are already developing in the jaws, but they cannot be seen. The incisors are the first to grow through the gums and this happens when the baby is about six months old. The first premolars appear at about fifteen months, and at three years the full milk dentition, totalling 20 teeth, is functional. This consists of two incisors, one canine, and two premolars on one side of each jaw. At intervals after this the molars erupt and the milk teeth come out and are replaced by a second set. This complete set, totalling 32 teeth, is called the **permanent dentition**. The details of this

Fig. 17:7 Comparison of the permanent and milk dentition in man.

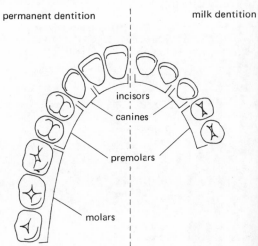

permanent dentition | milk dentition

incisors

canines

premolars

molars

174

can be summarised by a dental formula, which represents the teeth in the upper and lower jaws on one side.

$$I\frac{2}{2} \qquad C\frac{1}{1} \qquad P\frac{2}{2} \qquad M\frac{3}{3}$$ i.e. 16 teeth on each side

Present in milk dentition; replaced in permanent dentition.

Structure of teeth

Examine various kinds of teeth such as those of a sheep or a dog, including one tooth which has been cut vertically.

All mammalian teeth are embedded in sockets in the jaw (Fig. 17:8) and have the same essential structure. The main portion is composed of a bony material, the **dentine**, which is a living tissue. The part of the tooth which projects from the jaw is covered by a hard non-living secretion, the **enamel**. Where the tooth meets the gum, the enamel is replaced by cement which fixes the tooth firmly into the fibrous lining of the socket. Each tooth has one or more roots which are open at the base; these allow blood vessels and nerves to penetrate into the central part or **pulp cavity**. These roots almost close in most teeth, so preventing further growth, but in elephants' tusks, the incisors of rabbits and rodents, and the cheek teeth of herbivores, they remain open and allow growth to continue throughout life.

Fig. 17:8 Diagram of human teeth in longitudinal section: a) Incisor b) Molar.

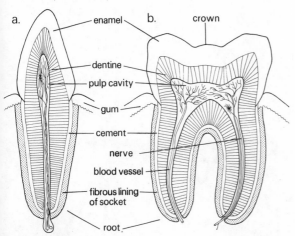

Other forms of dentition

Man is said to be an **omnivore** because he eats a great variety of food, both animal and vegetable. As we have seen, our dentition, consisting of all four kinds of teeth, is well adapted to cope with this wide range of foods. The **carnivores** which feed largely on meat, and the **herbivores** which feed on vegetation have characteristic dentitions which are especially suitable for these diets.

1. A carnivorous dentition

This is typical of lions, bears, otters, etc., but it is easier for you to study a cat or dog.

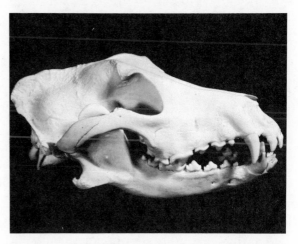

Fig. 17:9 Skull of dog showing a typical carnivorous dentition.

Examine a dog's skull. Note the small close-set incisors which are used for scraping small pieces of meat from the bone. They are also used for grooming the fur. Many species apparently derive pleasure from grooming each other.

Examine the very large canines which are used for gripping and may help in killing the prey. Why are they pointed in a backward direction? Why are they placed near the front of the mouth?

How does the shape of the cheek teeth differ from your own? The one on each side of the jaw which is larger than the others is called the **carnassial** tooth. When carnassial teeth come together they act like shears to cut the flesh away; they are also used for cracking bones.

The two jaws have an up-and-down movement with little sideways play. This makes the action more effective, as it enables the incisors to meet exactly and the carnassial teeth to slide accurately against each other, to help the shearing action.

The dental formula for a dog is:

$$\text{I}\,\frac{3}{3}\quad \text{C}\,\frac{1}{1}\quad \text{P}\,\frac{4}{4}\quad \text{M}\,\frac{2}{3}$$

Check this with your specimen. The fourth upper premolars and the first lower molars are the carnassial teeth.

2. A herbivorous dentition

Cattle, sheep, horses, antelopes and deer all have very typical herbivorous dentitions.

Examine a sheep's skull (Fig. 17:10) and see how it contrasts with that of the carnivore. Note how the incisors and canines are present only in the lower jaw. In life, these sharp teeth bite against a strong horny pad in the upper jaw; they are excellent for cutting tough grasses and other herbage.

The large gap between the front and back teeth is the **diastema**; it allows the tongue to manipulate the grass.

Note how all the cheek teeth are close together and form a large surface for grinding. When they first erupt they are more pointed, but quickly their crowns are worn away. As the enamel is harder than the dentine it wears more slowly and so projects to form the hard ridges. Look at those cheek teeth from the

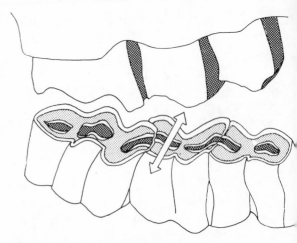

Fig. 17:11 Cheek teeth of sheep showing how the grinding surfaces fit together and move sideways.

side; note how the upper ones have a W pattern and the lower ones an M, so that they fit exactly.

When you watch a herbivore chewing, the sideways movement of the jaw is very marked; this allows the ridged surfaces to slide on each other and pulverise the grass, a process of great importance for effective digestion (Fig. 17:11).

The dental formula for a sheep is:

$$\text{I}\,\frac{0}{3}\quad \text{C}\,\frac{0}{1}\quad \text{P}\,\frac{3}{3}\quad \text{M}\,\frac{3}{3}$$

In other vegetarian animals such as voles, mice and rabbits the incisors have chisel edges which slide on each other; they are excellent for gnawing. As in sheep, a diastema is present, the cheek teeth are close together and their surfaces are flat and ridged for grinding. There are no canines. The dental formula for the rabbit is:

$$\text{I}\,\frac{2}{1}\quad \text{C}\,\frac{0}{0}\quad \text{P}\,\frac{3}{2}\quad \text{M}\,\frac{3}{3}$$

Fig. 17:12 Skull of rabbit—note the chisel-shaped incisors for gnawing.

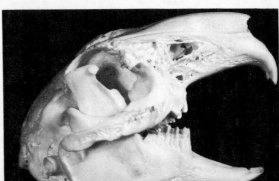

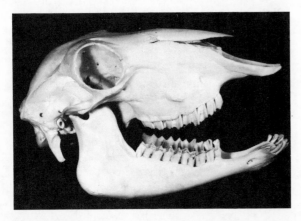

Fig. 17:10 Skull of sheep showing herbivorous dentition.

Care of the teeth

It has been said that man is unique in having three sets of dentition: milk, permanent and false! The need for false teeth and a lot of repair work could often be avoided if we treated our teeth better.

Dental decay is a major factor causing ill health. It starts when acid attacks the enamel, causing it to dissolve. The underlying dentine, being softer, then dissolves more quickly, allowing bacteria to enter. How is this acid formed? After a meal, food particles may lodge between the teeth or form a pulpy covering especially near the base. This is particularly the case after one has eaten sugary and starchy foods. If this food is not removed, bacteria act on it and break it down into acids which then attack the teeth. If we eat such things as chocolates, sweets and ice lollies throughout the day we provide a continuous supply of sugar for the bacteria to act upon, so our teeth are liable to decay quickly. This is one very important reason why dental decay has increased so alarmingly in recent years. Eskimos and many Africans were proverbial for having almost perfect teeth, but today when many of them have changed from their traditional diets to those of a Western style with more sugary foods, the incidence of dental decay has greatly increased. In Britain dental decay is so serious that dentists have even been doing research into the possibility of sealing the grooves of newly-erupted teeth with a hard plastic, but this research is at present in a very experimental stage.

The fact that false teeth are poor substitutes for natural ones is one more good reason why it is very important to treat teeth with the greatest care. Brushing teeth helps to remove both food and bacteria, and toothpaste is useful because it usually contains a mild alkali to neutralise the acid and a fine abrasive to clean the surface more effectively. Teeth should always be brushed in such a way that the gums are not pushed away from the teeth, as this exposes the weak spot below the enamel where the acids can more easily start their action.

Test the effectiveness of your particular brand of toothpaste for removing the bacterial film from teeth by carrying out the following test. You will need a toothbrush and some toothpaste. The method is to use erythrosin, a dye which stains the bacterial film red.

First rub some Vaseline over your lips to prevent them from being stained by the dye, then suck an erythrosin tablet. Examine your teeth in a mirror to see the extent of the staining on your teeth. Now brush your teeth thoroughly using the toothpaste and suck another tablet. The more effective the toothpaste, the less the teeth will stain the second time!

As well as brushing our teeth and avoiding eating sugary foods between meals, there are other ways of keeping teeth healthy. We have already seen that sufficient calcium should be present in the diet (p. 162), and that fluoridation of the water does much to harden the enamel and so keep the teeth healthy (p. 163). It is also important that we should have frequent, regular visits to a dentist for checking, as early treatment can often prevent the loss of a tooth. Babies and young children should be given hard foods to bite on as this helps to develop a good circulation to the teeth while they are growing, so aiding the conduction of the nutrients to them. Why is it good to end a meal with an apple or a raw carrot?

18

Making use of the food

The need for digestion

When we have eaten and swallowed our food it is still not inside our bodies in the strict sense of the word. We saw (p. 40) that earthworms were built up on a plan of a tube within a tube, the inner tube being the gut. The same applies to most of the higher animals. Before the food can be used it has to pass through the wall of the gut into the blood system which takes the food to the tissues which need it. Digestion is the process which makes absorption possible. Obviously solid foods have to be made soluble before they can pass through, but what about solutions? Do they need digesting too? It is not easy to experiment with a piece of living gut, but you could use instead a piece of cellulose Visking tubing which is rather similar to the gut regarding the substances which will pass through it. By filling this tubing with a solution of two foods, glucose sugar and starch, and placing the tubing in water, you could test the water to discover whether or not they are able to pass through the wall of the tubing. Proceed as follows:

Make up a solution of glucose and starch by adding 10 g of glucose to 100 cm^3 of 5% starch solution. Now take a piece of Visking tubing a little longer than a boiling tube and knot one end after wetting it. Pour the solution into it and tie the end with cotton (Fig. 18:1). Rinse the outside of the tube with cold water to remove any solution which may have been spilt and place it in a boiling tube containing enough distilled water to surround the tube. Now remove with a syringe a little of the surrounding water, place a drop on a tile and test for starch by adding iodine/potassium iodide solution; boil the rest with Benedict's solution to test for glucose. Repeat the tests after five and ten minutes, and again after half an hour. What do you conclude from your results, remembering that glucose has small molecules and starch very large ones?

Membranes such as the cellulose one used as a model gut, and those composed of living cells (like those lining the gut) allow some molecules through and not others. Digestion involves the breaking down of molecules until they are small enough to pass through. This is done in the gut by means of digestive enzymes, protein substances which act as

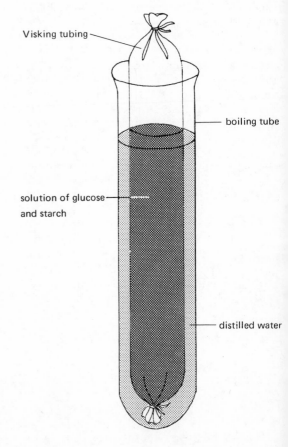

Visking tubing

boiling tube

solution of glucose and starch

distilled water

Fig. 18:1 Experiment to find out whether starch and glucose will pass through an artificial membrane.

catalysts and greatly speed up the reactions. We are able to produce many enzymes from special glands which act on the three main categories of complex food that we eat—carbohydrates, proteins and fats—and change them into soluble substances with small enough molecules to pass through the membranes of the gut. As a result, digestible carbohydrates are turned into simple **sugars**, fats into **fatty acids** and **glycerol** and proteins into **amino acids**.

Investigating a digestive process

Before describing how and where these reactions are carried out we will try out a digestive process for ourselves.

Is starch digested with the help of saliva?

When we eat starchy foods such as bread or cereals we mix them up in our mouth with saliva. Does the saliva contain an enzyme which helps to digest starch? When starch is digested, it is turned into a reducing sugar, so to find out if saliva is involved in this process, you would have to mix some starch with saliva and carry out tests to see if the starch was removed and reducing sugar appeared. But first you would have to be sure that no reducing sugar was present at the start. Proceed as follows:

Prepare a water bath by heating a beaker of water to 37°C (body temperature).

Rinse out your mouth thoroughly with water to remove any sugar. Take a mouthful of distilled water, swill it round for half a minute so as to mix it with the saliva and pass it into a boiling tube.

Test 2 cm³ of the saliva solution by boiling with 2 cm³ Benedict's solution to check that no sugar is present. (As a control also boil 2 cm³ Benedict's solution with water.)

Place 2 cm³ of saliva in a test tube and 2 cm³ of a 1% solution of pure starch into a second tube and place both in the water bath.

Put a series of drops of iodine/potassium iodide solution on to a white tile.

Now mix the saliva and starch in one of the tubes, stir with a clean glass rod and immediately put a drop of the mixture on to one drop of iodine on the tile. Stir it and note the colour.

Clean the glass rod and repeat at ½-minute intervals, noting any change of colour, and proceed until no further colour change takes place.

Finally add 2 cm³ of Benedict's solution to the remaining mixture and boil. Is reducing sugar present? What is your final conclusion from this experiment?

What is the effect of temperature on this reaction?

You could find out the effect of temperature on this enzyme action by modifying the last experiment. This would involve taking equal quantities of starch and saliva and finding out how long it takes to turn the starch into reducing sugar when the mixture is kept at different temperatures. At the same time you could see the effect of boiling the saliva.

Take six test tubes and label them A–F. Add 2 cm³ of 1% starch solution to A, B and C and 2 cm³ of saliva to D, E and F.

Boil the saliva in F.

Place A and D in a beaker of water kept at 20°C and the other four test tubes in a beaker of water kept at 37°C.

After a few minutes (why?), add the contents of A to D, B to E and C to F, and take the time when each is mixed. Record the time taken for the mixtures to give no colour with iodine on a tile. What do you conclude from your results?

What effect had the boiling on the saliva?

A similar kind of experiment can be used to determine the effects of acid and alkali on this enzyme action. Think out the experimental procedure you could use and the precautions you would take using dilute hydrochloric acid and sodium bicarbonate as the acid and alkali.

Digestive enzymes

We have now considered one enzyme action. In the gut there are many of these digestive catalysts which help to break down the carbohydrates, fats and proteins into soluble and absorbable products. They can be classified into three groups:

1. **Amylases** which act on carbohydrates and turn them into simple hexose sugars (those

with six carbon atoms), e.g. glucose, fructose, lactose.

2. **Lipases** which act on fats and turn them into fatty acids and glycerol.

3. **Proteases** which act on proteins and turn them into various amino acids.

Many of these reactions involve several steps, each one being catalysed by an enzyme. For example, the enzyme that you investigated in saliva, **ptyalin**, converts starch to maltose sugar. In the small intestine another enzyme, **maltase**, converts the maltose to glucose.

$$\text{starch} \xrightarrow{\text{ptyalin}} \underset{C_{12}H_{22}O_{11}}{\text{maltose sugar}} \xrightarrow{\text{maltase}} \underset{C_6H_{12}O_6}{\text{glucose}}$$

Note that the name of the sugar ends in *ose* and the enzyme which acts on it in *ase*.

Effect of pH and temperature on enzyme action

You should have discovered from your experiment on ptyalin that the enzyme reaction varied with temperature. Fig. 18:2 shows how the rate of the reaction varies and how the enzyme is inactivated if heated too much. From this graph determine the optimum and maximum temperatures for enzyme action and compare the optimum with your own body temperature.

Each enzyme also has its special requirements, some working best in an acid, others in an alkaline medium. As the pH (degree of acidity or alkalinity) alters considerably during the passage of food through the gut, different

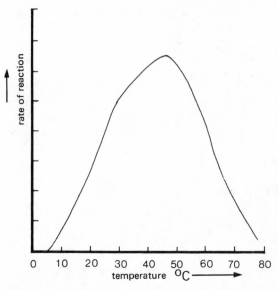

Fig. 18:2 Graph showing how the rate of enzyme action varies with the temperature.

enzymes are needed in various parts which can work effectively under these particular conditions. Thus the gastric enzymes act in a strongly acid medium and those in the duodenum in a progressively less acid one.

Alimentary canal of man

We can now study the digestive system of man and see how it is adapted to deal with food. It is best considered as a long tube open at both ends (mouth and anus) which is modified in different places to deal with the preparation, digestion and absorption of the food and the elimination of the remaining material or

Fig. 18:3 Diagram showing how the alimentary canal is a much modified tube.

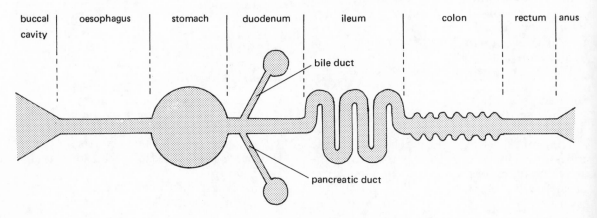

180

faeces. The tube is very long because digestion is a slow process and absorption is most effective where there is a large surface area in contact with the digested food.

Compare Fig. 18:3 with the more detailed diagrams of the alimentary canal of man and of the rat (Figs. 18:5 and 18:6). You should also examine a dissection of this system in a rat to see what it really looks like. What differences do you notice between these systems in man and rat?

Let us consider in more detail how each part of the system in man carries out its particular part in the process.

Buccal cavity (Fig.18:4)

Here the food is broken up by the teeth, a process called **mastication**. This not only reduces the food to a convenient size for

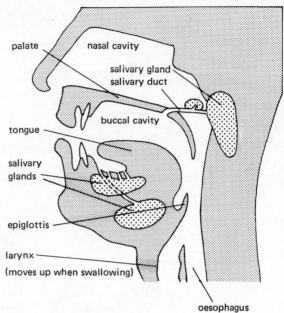

Fig. 18:4 Vertical section through the buccal cavity region of man.

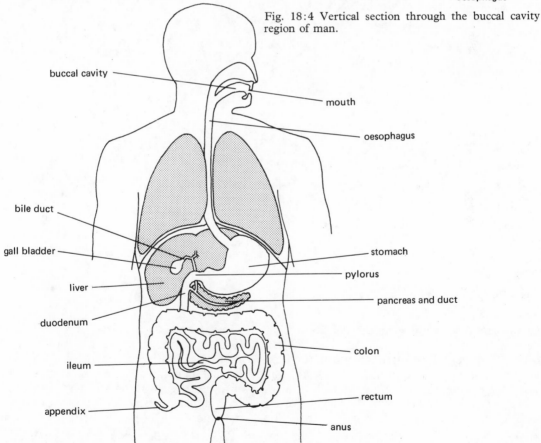

Fig. 18:5 Alimentary canal of man.

181

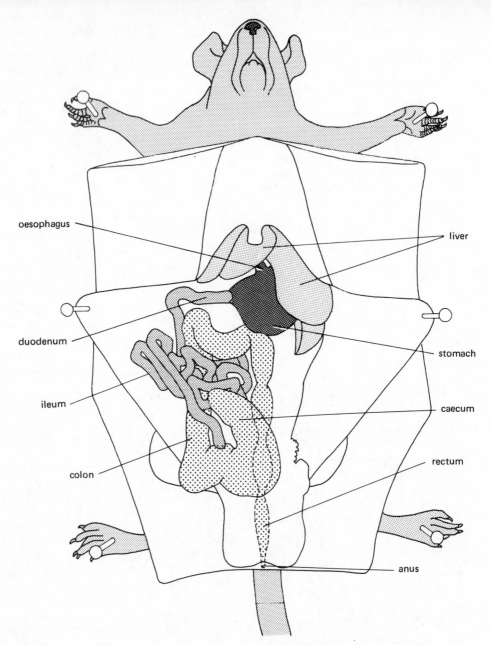

oesophagus

liver

duodenum

stomach

ileum

caecum

colon

rectum

anus

Fig. 18:6 Rat dissected to show the alimentary canal.

swallowing, but it also increases the surface area of the food so that enzymes can digest it more effectively. During mastication, the food is moved about by the tongue so that all portions are broken up and mixed with the saliva which helps to lubricate and bind it together into a ball or **bolus** for swallowing. The saliva comes from 3 pairs of salivary glands; it contains the enzyme ptyalin, which we have already seen turns starch into maltose sugar.

Oesophagus

This is a muscular tube which conveys the food through neck and thorax regions to the stomach; it lies just behind the windpipe.

When we swallow slippery food, after the tongue has pushed it to the back of the throat, we might have the impression that it falls down the tube, but in fact the food is squeezed down by a process called **peristalsis**. The walls of the oesophagus contain two layers of muscle arranged in a circular and longitudinal manner (Fig. 18:7). When the circular muscle contracts just *behind* the bolus, the food is squeezed downwards; this occurs rhythmically, rather like squeezing a marble down a piece of rubber tubing with your fingers. When the longitudinal muscles contract *in front* of the bolus they cause the tube to widen, thus aiding the process. Peristalsis is so effective that even liquids can be swallowed when we are standing on our heads. Think what happens when a giraffe drinks; it is uphill all the way! You can tell how long it takes you to swallow if the food is very cold, e.g. ice cream, as you can feel it when it reaches the stomach.

Stomach

This is a muscular sac with extendable walls primarily concerned with the early stages of protein digestion. The only absorption of food that takes place here is of substances with small molecules which need no digestion, such as glucose and alcohol. With the breakdown of tissue membranes, fats are also released into the stomach cavity, but no digestion of these takes place.

Both ends of the stomach are closed most of the time by sphincter muscles; these contain fibres arranged in a circular manner which

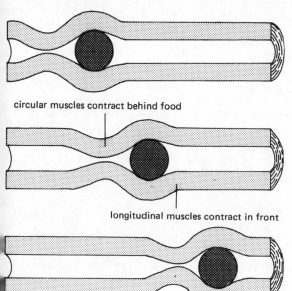

circular muscles contract behind food

longitudinal muscles contract in front

Fig. 18:7 Diagram showing the principle of peristalsis.

when contracted close the aperture. The most important of these is the **pylorus** which guards the entrance to the small intestine. Periodic waves of peristaltic action of the stomach wall (about every 20 seconds) cause the food to be thoroughly mixed up with the gastric juices. The food does not escape through the pylorus until it is in a semi-liquid state called **chyme**. Even when the stomach is empty muscular contractions may still occur; this may cause 'tummy rumblings' and sometimes a feeling of hunger.

Gastric juices are secreted on to the food from glands in the stomach wall, some producing hydrochloric acid and others, proteases. Of these, pepsin starts protein digestion and rennin, which is particularly important in babies, acts on the protein in milk (caseinogen) and changes it into casein, thus causing the milk to clot. You may wonder why the acid and pepsin do not digest the living tissues lining the stomach as these are largely composed of protein. This action is prevented by the secretion of mucus which forms a protective layer over the surface of both the stomach and subsequent regions of the gut. If for some reason this protective layer becomes less efficient, ulcers may result (p. 187).

The hydrochloric acid not only provides an acid medium for these enzyme actions, but it also kills bacteria and possibly parasite eggs which may enter with the food. In carnivores such as dogs and hyaenas the acid will dissolve the bones they eat.

The rate at which the stomach empties depends on the type of food being digested; a light meal in semi-liquid form may pass through the pylorus within an hour, but a heavy meal with a lot of protein and fat may remain in the stomach for three or four hours.

The small intestine

This is a long tube which varies in length according to its degree of contraction; in life it is probably about five metres long. Its two functions are to finish digestion and absorb the products. The first loop of the small intestine is the **duodenum**; it receives the bile and pancreatic ducts and is largely concerned with digestion. The very much longer hind region of the small intestine is the **ileum** which deals mainly with absorption.

When the pylorus opens, and the acid chyme from the stomach is squirted into the duodenum, three things happen:

1. The gall bladder in the liver contracts and the bile which is stored there passes down the bile duct and so into the duodenum.

2. The walls of the duodenum are stimulated to secrete into the blood a hormone (chemical messenger) called **secretin**, which causes the pancreas to discharge pancreatic juice down the duct and so on to the food.

3. Glands in the wall of the duodenum discharge a digestive juice, **succus entericus**, on to the food. These three juices work together to bring about complete digestion. All are alkaline, and so they help to reduce the acidity of the chyme, thus providing a suitable medium for the enzymes to work in.

Bile is a green fluid which is secreted by the liver cells and stored in the gall bladder. It contains bile salts which give it the colour, and sodium bicarbonate which makes it alkaline. It contains no enzymes. The bile salts act like a detergent in breaking up any fat in the food into microscopic droplets, a process called **emulsification**. This process helps digestion as it increases the surface area of the fat for the enzymes to work on.

Pancreatic juice is secreted by the pancreas which is a gland lying below the stomach; its duct joins the bile duct before entering the duodenum. It contains a protease, an amylase and a lipase.

Succus entericus, secreted by glands in the intestine wall, contains a number of enzymes which complete the digestion of carbohydrates into various hexose sugars and finishes the breakdown of proteins into soluble amino acids. A summary of the digestive process is shown in the table below.

Absorption of the products of digestion

Absorption of the end products of digestion occurs mainly in the ileum. The efficiency of the process is increased because:

1. The products pass along it slowly, perhaps taking as long as three or four hours.

2. Its walls are not only folded internally to form ridges, but its lining is thrown into millions of finger-like processes called **villi** which greatly increase the surface area for absorption.

Summary of Digestive Processes (enzymes in bold type)

Region	Name of secretion	Where produced	Contents of secretion		Action
Buccal cavity	Saliva	3 pairs of salivary glands	(i)	**Ptyalin** (salivary amylase)	Starch → maltose. Lubricates food.
			(ii)	Water	
			(iii)	Mucin	
Stomach	Gastric juice	Glands in stomach wall	(i)	**Pepsin**	Proteins → intermediate products.*
			(ii)	**Rennin**	Clots milk (caseinogen → casein).
			(iii)	Hydrochloric acid	Provides acid medium for pepsin; kills bacteria.
Small intestine	Bile	Liver		Bile salts (no enzymes)	Emulsifies fats.
	Pancreatic juice	Pancreas	(i)	**Trypsin**	Proteins → intermediate products* and some amino acids.
			(ii)	**Amylase**	Starch → maltose.
			(iii)	**Lipase**	Fats → fatty acids and glycerol.
	Intestinal juice (succus entericus)	Duodenum	(i)	**Sucrase**	Sucrose → glucose and fructose.
			(ii)	**Maltase**	Maltose → glucose.
			(iii)	**Peptidases**	Intermediate products formed from protein* → amino acids.

*Protein digestion occurs in stages whereby the molecule becomes progressively smaller. These intermediate products, e.g. peptones and polypeptides, are finally broken down into amino acids.

3. The rhythmic contractions of the wall ensures that the fluid which is in contact with the absorptive surface is constantly changing.

Each villus (Fig. 18:8) contains a network of capillaries just below its surface so that hexose sugars and amino acids can quickly pass from the gut into the blood stream. These capillaries join up to form the large portal vein which takes these substances to the liver. The fatty acids and glycerol are mainly absorbed into the lymphatic system, a branch of which —a **lacteal**—is present in each villus; these products eventually reach the blood system via a duct in the neck.

The large intestine

In man, this consists of the colon and rectum. By the time the contents of the gut reach the colon the main products of digestion have been absorbed. The residue consists largely of water and solid material such as vegetable fibres and cellulose for the digestion of which there are no appropriate enzymes. In addition, bacteria are present in astronomical numbers. The chief function of the colon is to absorb most of the remaining water, leaving the solid material as faeces which pass into the rectum

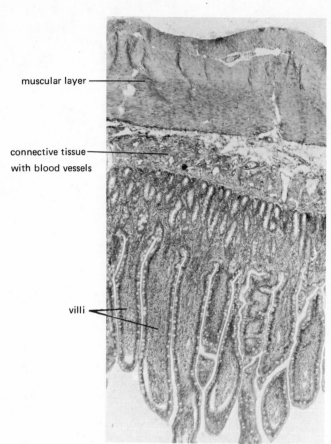

muscular layer

connective tissue with blood vessels

villi

Fig. 18:9 (above) Photomicrograph of a transverse section through the ileum of a cat.

Fig. 18:10 (below) Photomicrograph of a similar section with the blood vessels injected to show the circulation.

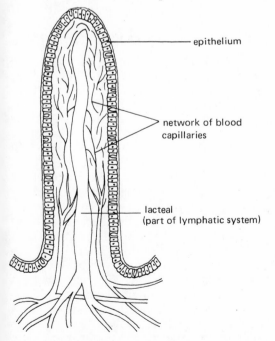

epithelium

network of blood capillaries

lacteal (part of lymphatic system)

Fig. 18:8 Diagram of a villus much enlarged.

185

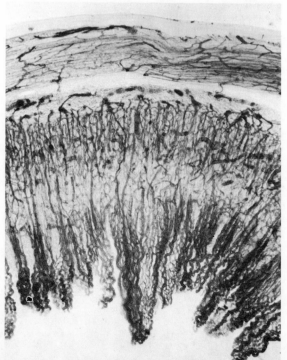

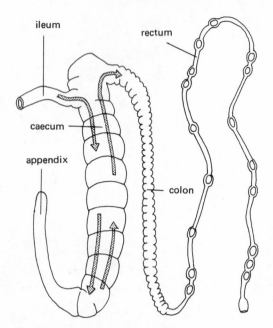

ileum

rectum

caecum

appendix

colon

Fig. 18:11 Large intestine of rabbit.

and are periodically eliminated through the anus.

In many mammals, especially herbivores such as rabbits, sheep and cattle, the large intestine also includes a large blind out-growth, the **caecum** which ends in the **appendix** (Fig. 18:11). In man these are greatly reduced in size and have no digestive function.

The liver

As already indicated, hexose sugars are trans-ported from the villi via the portal vein to the liver. An important function of the liver is to store this sugar temporarily in its cells in the form of solid **glycogen**, a more complex carbohydrate rather like starch.

The liver also regulates the amount of sugar that passes through it into the general circula-tion. Otherwise the level of sugar in the blood would rise steeply after the digestion of each meal. What osmotic effect would this have on the body cells? As sugar is constantly being removed from the blood by the body cells to replace what is used up in respiration, more sugar is needed to take its place. The liver supplies this need by converting some of its stored glycogen back into sugar and releasing it into the blood stream.

The amino acids cannot be stored in the body but travel round the general circulation and are selectively absorbed by the cells which need them for building up their pro-teins. Any surplus amino acids are broken down by the liver into a carbohydrate which is stored as glycogen and urea which is passed to the kidneys for excretion. This process is called **deamination**.

Fatty acids and glycerol eventually arrive in the liver via the lymphatic system and the blood stream. Here the fatty acids are broken down into smaller units. Some are converted by the liver cells into carbohydrates while others are reconverted into fats by cells in various parts of the body, especially those under the skin and in the connective tissues which surround the kidneys and other organs. This fat acts as a reserve of fuel which can be oxidised in respiration to liberate energy.

Additional functions of the liver include the secretion of bile (p. 184), the storage of iron for the manufacture of more haemoglobin, the formation of fibrinogen, a protein used in blood clotting, and the conversion of certain poisons, which may enter the bloodstream as a result of bacterial action in the colon, into harmless substances—a process called **detoxi-cation**.

The digestive process in herbivores

Because herbivores consume vegetable ma-terial they take in great quantities of cellulose, but like all mammals they are unable to digest it as no enzyme is present that will break it up. However, some of the bacteria living in various parts of the guts of herbivores are able to do this. In the rabbit, these bacteria occur in vast numbers in the large caecum, but in species which 'chew the cud' such as sheep and cattle, they largely occur in their specialised four-chambered stomachs.

In sheep and cattle, the herbage is first chewed and mixed with saliva. It is then swallowed and passed first to the **rumen** (Fig. 18:12) and then to the **reticulum**, the first two chambers of the stomach. In both these chambers astronomical numbers of bac-teria start the breakdown of cellulose. This action continues when the food, or **cud**, is re-gurgitated into the mouth and subjected to further pulverisation by the teeth. When the

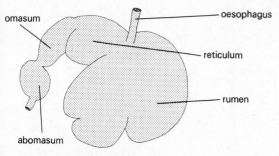

Fig. 18:12 Sheep's stomach.

cud is swallowed once more it is directed towards the third chamber, the **omasum**, where it is churned up and some water is extracted from it. It then enters the fourth chamber, the **abomasum**, which is the part that corresponds to the stomach of other mammals. Here hydrochloric acid and proteases are secreted on to the food and the bacteria are killed by the acid. The products of cellulose digestion by the bacteria are mainly fatty acids which are absorbed by the blood and become the main source of energy for the sheep or cattle.

Some mammals, such as rabbits, have the strange habit of eating their own faeces. When the food passes through the gut the first time it is only partly digested and the faeces which are formed are moist; in wild rabbits these are voided during the day when below ground. These faeces are then eaten at once and further digestion of their contents takes place. During their second passage most of the water is extracted from them so these faeces when voided are dry; these are the pellets you commonly see in grassy places in districts where rabbits are common. Can you see any possible connection between this habit and the ability of rabbits to live in very dry places?

Health aspects of the alimentary canal

The human alimentary canal usually functions remarkably well considering how badly we treat it on occasions! Sometimes it rebels, and we either feel sick or have indigestion.

Vomiting is the body's method of ridding itself of unwanted or harmful substances from the stomach. The peristaltic movements of the stomach and oesophagus reverse their normal direction and the food is expelled. There are many causes of vomiting, but one of the most common is over-eating, especially when the food contains a high proportion of fat. Vomiting also occurs when we eat something very indigestible or poisonous.

When we feel 'bilious' or 'liverish', it is often the result of having eaten 'rich' meals over several days. The liver is unable to cope with the excessive fat and we get a feeling of nausea and sometimes a headache.

Indigestion is a general term used when there is difficulty in digesting food. Healthy people can usually avoid indigestion by a) having simple, well-balanced meals, b) eating them in a leisurely manner, c) thoroughly masticating the food, and d) avoiding taking violent exercise soon afterwards. We can learn a lot from the animals which after a meal have a good sleep!

A more serious form of indigestion is caused by stomach and duodenal ulcers. These conditions occur more often in people who may be described as hurried or worried. Thus ulcers occur more often in busy people who get into the habit of hurrying over meals and rushing from one activity to another without sufficient rests—such as doctors, schoolmasters, members of parliament, stockbrokers and business executives. Those who are able to relax, who are not continually tensed up, and who live at a slower pace, seldom get ulcers.

For good health it is necessary to empty the bowels regularly. If the food residues remain in the colon for too long, the bacteria present have more time to produce harmful substances which may be absorbed by the blood. Constipation can often be avoided by having plenty of roughage in the diet (p. 165).

19

Bacteria and viruses

We discussed (p. 145) the different methods of nutrition used by living organisms and you will recall that the basic method is holophytic—the typical method employed by green plants. All other organisms, because they lack chlorophyll, are dependent for their food upon the complex substances which green plants have made. Animals typically use the holozoic method of nutrition, taking their food into their bodies and digesting it before it is absorbed. Other organisms, such as *Mucor* (p. 42) and many other fungi, feed on dead organic matter by the saprophytic method by secreting digestive fluids on to the food and absorbing the soluble products. By contrast, parasites such as *Pythium* (p: 43), the malarial parasite (p. 71) and tapeworms (p. 36), feed on the living tissues of other organisms or on the food eaten by their hosts.

We are now going to study bacteria, the great majority of which feed either as saprophytes or parasites. Many of those which feed saprophytically are very beneficial and, like many fungi, they cause the decay of dead material and thus return essential nutrients to the soil. However, those which act as parasites attack living organisms, including man, and thereby cause disease.

Bacteria are so important to man that in this chapter we shall also study their general biology and some of the ways in which they affect our lives.

BACTERIA

These cannot be seen with the naked eye, but the largest are visible under the lower power of a microscope. Most come into the size range $0.5 \ \mu m$–$8.0 \ \mu m$ where $1 \ \mu m = 1$ micrometre (0.001 mm). If the average size of a bacterium is $1 \ \mu m$ you could calculate how many there would be if they were placed in line on a 10 cm ruler.

With very few exceptions, bacteria have no chlorophyll and, like fungi, their surrounding wall is made of protein and fatty substances, not cellulose as in normal plants. They are non-cellular and the majority have no nuclei, although nuclear material is scattered within the cytoplasm.

They are classified into three groups according to their shape (Fig.19:1):

1. **Cocci** (sing. coccus), which are spherical. Sometimes they are grouped together in pairs (diplococci), others in groups of varying sizes (staphylococci) and others in strings (streptococci).
2. **Bacilli** (sing. bacillus) which are rod-shaped. Sometimes these have long projecting cytoplasmic threads called flagella which are used in movement.

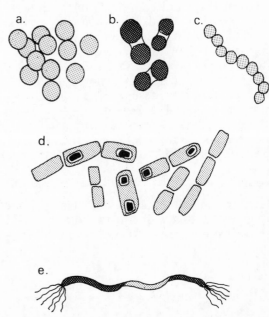

Fig. 19:1 Types of bacteria: a) Staphylococci. b) Diplococci. c) Streptococci. d) Bacilli (some with spores). e) Spirillum with flagella.

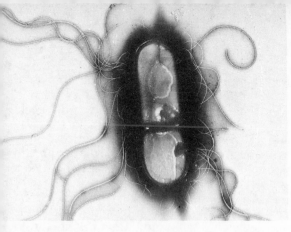

Fig. 19:2 Electronmicrograph of a bacillus with flagella. × 10,000.

3. **Spirilla** (sing. spirillum) which are twisted like a corkscrew, the number of twists varying from one to many. Some of these also have flagella at their ends.

BACTERIA AS SAPROPHYTES

If some chopped-up meat or a dead earthworm is left in a beaker of water, the water soon becomes cloudy, a scum forms on the top and it starts to smell. This is because saprophytic bacteria are causing the decay of the meat, the proteins being broken down into simpler nitrogenous compounds which are volatile and therefore smell. During this process of decay the bacteria absorb nutrients, grow and multiply very rapidly. Reproduction is a simple process of binary fission which can take place roughly every half-hour at room temperature. So, if there was only one bacterium to begin with, after 24 hours the number produced would have 14 noughts in it! This may be surprising, as after an hour there would only be 4, after $1\frac{1}{2}$ hours only 8 and after 2 hours 16. When large numbers of bacteria are produced they tend to collect near the surface where there is more oxygen; here they produce a lot of mucilage and it is this which causes the scum on top of the beaker.

Examine under the high power of the microscope a small drop of water in which something has decayed. If a microscope with an oil immersion lens is available, this would be better still. Focus very carefully and look out for bacteria of different shapes in a state of movement. The cocci joggle about in a haphazard way because they are being bombarded by molecules in the water (Brownian movement). Bacilli may move in a particular direction as a result of their flagella (not visible) while spirilla can move quite fast like animated springs.

Although some bacteria have to have oxygen for the liberation of energy from food, others can do without it by respiring anaerobically and others, rather strangely, cannot exist if oxygen is present. Many species form spores within their bodies (Fig. 19:1d); these have hard resistant coats which allow them to remain alive for long periods under difficult conditions. When better conditions return the coat breaks down and active life is resumed.

Where do bacteria come from?

In the Middle Ages it was believed that many living organisms arose spontaneously from non-living things. Thus it was thought that frogs were derived from the mud of ponds, maggots from bad meat and mice from dirty clothes left in a cupboard! As a result of more critical observations and experiments these conclusions were shown to be false, but the idea of **spontaneous generation** persisted for many of the smaller organisms and when bacteria were discovered it was widely believed that they arose from decaying material.

It was in 1861 that **Louis Pasteur** (1822–95), the famous French scientist, finally proved that spontaneous generation of bacteria did not occur and that food went bad because bacteria were carried to it via the air. The experiment he performed to prove this is a classical example of how a scientific experiment should be done using a control.

For this experiment he used a clear nutrient broth which, when exposed to the air, would normally have gone bad after a few days. First he had to kill off any bacteria already in the broth. This he could do quite easily by boiling it; but how could he then allow air to come in contact with the broth without bacteria settling on it as well? He thought out a most ingenious method for doing this. He put the broth in a flask with a long thin neck and then bent the neck into an S shape in a flame (Fig. 19:4). He boiled the broth to kill off any bacteria in it. Now, if the flask was cooled, he knew that the vapour would condense inside and due to the reduction in pressure air would be drawn into the flask. So he reckoned that if

Fig. 19:3 Louis Pasteur (1822–1895).

Fig. 19:4 Pasteur's flask.

the cooling was done slowly, the air would pass in but the bacteria, being heavier than air, would not pass beyond the bend in the neck. He prepared a number of flasks in this way and the experiment worked perfectly; however long the flasks were left, the broth did not go bad. But to make quite sure, he used a control experiment by cutting off the neck of one of the flasks so that air could enter *from above*. He reasoned that if bacteria were in the air they would fall on to the broth. The broth went bad within a few days!

The success of these experiments depended on Pasteur's ability to kill off the bacteria which were in the broth to begin with by boiling, a process called *sterilisation*. Today heat is still used to sterilise apparatus and the media for culturing bacteria.

Growing bacterial cultures

Microbiologists today usually culture bacteria on nutrient **agar**. This is a jelly-like material extracted from seaweed to which various food substances are added, such as beef extract, to suit the requirements of the bacteria to be

Fig. 19:5 Method of inoculating an agar plate using a nichrome loop.

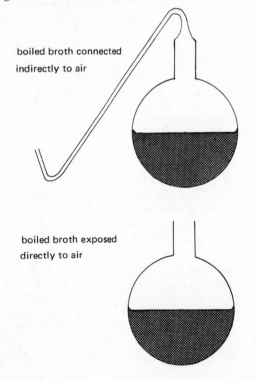

boiled broth connected indirectly to air

boiled broth exposed directly to air

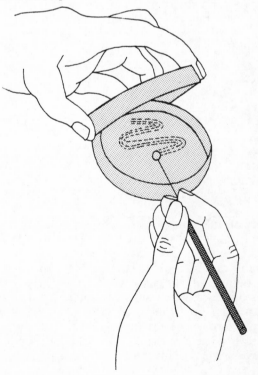

190

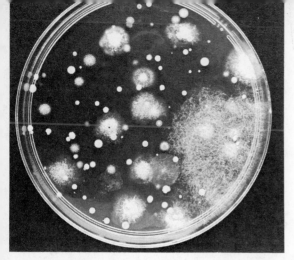

Fig. 19:6 Cultures growing on an agar plate after exposure to air for 30 seconds. The smaller colonies are bacteria, the larger, fungi.

and when cool transfer a drop on to the agar, making a regular pattern (why?).
3. To test tap water: transfer a drop as in 2.
4. To test milk: as in 2.
5. To test your skin: make some fingerprints on the surface of the agar by touching it lightly.
6. To test your breath: hold the plate near your mouth and cough into it several times.
7. Inoculate from any other source, e.g. a scraping from your teeth or from under your finger nails.
8. Leave untouched to act as a control.
Place all dishes in an incubator at 35–37°C and examine them after about 48 hours.

cultured. (Fungi may be cultured in the same way, a good medium being agar to which potato extract and glucose are added.)

The nutrient agar is dissolved in water, sterilised by boiling it under pressure in an autoclave (rather like a pressure-cooker) and then poured into sterilised Petri dishes where it sets on cooling to form a **plate.** The lid of the Petri dish is loose fitting and so allows air to enter, but bacteria are unable to do so—why is this? Tablets are commercially available which contain the ingredients for various types of medium and provide a quick alternative method for making up nutrient agar.

To culture bacteria a loop of nichrome wire is first held in a flame until red hot to sterilise it, and when cooled, dipped into the medium to be tested, e.g. a drop of milk. The lid of the Petri dish is then raised slightly and the loop stroked gently over the surface of the agar (Fig. 19:5); the lid is then quickly replaced. The Petri dish is placed in an incubator at 37°C for 48 hours. By this time any bacteria will have multiplied sufficiently to be visible as streaks where the loop touched the surface.

Working in groups, you can now use this method to find out how widespread bacteria are. Each group will need eight sterilised Petri dishes containing sterilised nutrient agar. Label them 1–8 with a chinagraph pencil and treat them as follows:
1. To test for bacteria in the air: take the lid off and expose for 15 minutes.
2. To test soil: shake up a little soil in distilled water; sterilise the loop in a flame

If the plates were sterilised successfully No. 8 should have remained clear. Did you find bacterial colonies in all the others or only some? In some, you may have grown fungus colonies as well; they may be recognised by their hyphae (p. 41). If there is any particular colony you wish to grow as a pure culture you can do this by sub-culturing. To do this you touch the colony with the sterilised loop and stroke it over the surface of sterilised agar in another dish. If a pure culture is not obtained first time the process can be repeated.

Most of the bacteria grown will have been saprophytes which cause decay, but some could be disease causing, i.e. **pathogenic.** Hence it is necessary to destroy the cultures when they are finished with by putting the Petri dishes, lids and all, into a bowl of strong disinfectant such as lysol.

Let us now consider some of the important implications of the experiments you have carried out, first of all those concerned with decay.

Bacteria and food

You probably established that air contains bacteria, so if food is left exposed to the air bacteria will settle on it and cause it to decay. If the food is damp and the temperature warm it will decay all the more quickly. This is a great problem in the tropics and probably explains why traditionally so many curry powders, herbs and spices were used in preparing food, as they masked the disagreeable flavour of food which was going bad. Today we can prevent food from decaying in many

191

ways, so allowing it to be stored for almost indefinite periods and transported all over the world. The more important methods are:

1. Refrigeration

This does not kill the bacteria, but the lower the temperature, the more it prevents them from multiplying and the slower their decaying action becomes.

Domestic refrigerators are usually kept at temperatures just above 0°C. This is low enough to reduce bacterial activity to an extremely low rate and at the same time to reduce the metabolic and transpiration rates of living food such as fruit and green vegetables so that they keep fresh for longer periods. Food does not remain fresh indefinitely at these temperatures and when it is taken out of the refrigerator bacterial activity will be resumed.

The use of a deep freeze is more efficient as the temperature is much lower and all metabolic activity is stopped, so food keeps indefinitely. By this method fresh meat, fish, fruit and vegetables can be 'put down' when supplies are plentiful and cheaper, and used when required.

In the food industry, refrigerated lorries and ships are used to overcome the problem of transporting perishable food over long distances.

The biggest refrigerator in the world is the Antarctic continent. You may have read that food left by Captain Scott during his famous expedition to the South Pole in 1912 was discovered many years afterwards still in perfect condition. It has been suggested that the Antarctic could become the world's storehouse for surplus food. Can you think of reasons for and against this idea?

Another example of natural refrigeration is when animals become embedded in ice. In Siberia mammoths which became extinct many hundreds of years ago have been discovered with the meat still in an edible condition!

2. Sterilisation by heat treatment

Several methods are used, all based on the facts established by Pasteur:

a) *Canning*. The food is subjected to high-temperature cooking, placed in sterilised tins, reheated under pressure and sealed when still hot. In this way bacteria in the food are killed and no others can enter. Bottling works on the same principle. These methods are very useful for meat, fish, fruit and vegetables.

If bacteria are not destroyed they give out carbon dioxide during their respiration and this causes the tin to bulge. The contents should on no account be eaten if this occurs as the bacteria will probably produce toxins. Also, beware of tins which are badly rusted as these could be contaminated.

b) *Pasteurisation of milk*. You should have found that milk contained plenty of bacteria; this is not surprising, as milk is an excellent food for micro-organisms as well as us! Most of the bacteria it contains will merely cause it to go sour, but pathogenic forms may also be carried in milk and multiply quickly. Pasteurisation is a method used for killing off the majority of bacteria, including those disease-producing forms, without spoiling the flavour of the milk by boiling it. The usual method is to heat the milk to 72°C for 15 seconds, quickly cool it to 12°C and put it into sterilised bottles which are capped at once. Some decay bacteria may survive this treatment, so it will go sour, but not so quickly as untreated milk. Another method is to heat the milk to a very high temperature (135°C) for one second and immediately put it into containers and seal. The milk keeps fresh for much longer.

3. Osmotic methods

The principle is that bacteria cannot survive in an active state in a solution of high osmotic strength as water is drawn out of them. Sugar and salt in high concentration have this effect, so that honey, jams and salted meat, fish or vegetables do not go bad. The sun-drying of various kinds of grapes to produce raisins, sultanas and currants has the same effect, as the drying concentrates the sugar solutions in their cells. The salting and sun-drying of fish which is practised in the tropics on a large scale works on the same principle (Fig. 19:7).

4. Dehydration

The method is to remove so much water from the food that bacterial action becomes negligible. This technique is suitable for milk, eggs, vegetables such as potatoes and many cereal products. After drying, it is necessary to keep the food in waterproof containers. A

Fig. 19:7 Fish being sun-dried, Wanseko, Uganda.

great advantage of this method is that it reduces the weight and bulk of the product, making transport cheaper and easier. In tropical countries where refrigeration is difficult, liquid milk may 'go off' a few hours after milking. Under these conditions the use of dried milk is of the greatest importance.

5. Chemical methods

Some foods are preserved by adding chemicals which kill bacteria but are considered to be harmless to man in the quantities used. This is the least satisfactory method as certain preservatives used in the past have been found later to be harmful. Strict regulations are now imposed to prevent the improper use of such preservatives. Substances still commonly used include benzoic acid and sulphur dioxide.

The smoking of fish is basically a chemical process as the smoke contains substances which are poisonous to bacteria and these become impregnated in the outer layers which are also dried and hardened in the process.

Pickling food in vinegar is another chemical method; it is successful because the acid kills the bacteria.

6. Irradiation

This modern and very effective method is to package and seal the food first and then irradiate it using radio-active cobalt; this kills all organisms which are present.

Bacteria and soil

Your experiment with a soil sample will have demonstrated that soil is teeming with bacteria. In fact, a pinch of soil will probably contain more bacteria than there are people in Britain. Most of these bacteria bring about the decay of organic matter in the soil. They are essential for life on our planet as they play a vital part in recycling the substances made by plants and animals so that they can be used once more.

Bacteria and sewage

Fresh water, sea water, well water and even tap water all contain bacteria, hence any organic matter present decays just as it does in soil. This can be a big problem, for example, when large quantities of raw sewage are passed into lakes and rivers. In this case so much bacterial action takes place that nearly all the oxygen dissolved in the water is used up and other living things die in consequence. Another problem arises when large quantities of raw sewage are passed directly into the sea because it happens to be the cheapest method of disposal; this makes the water foul, and possibly dangerous to bathers, at many popular holiday resorts.

Fig. 19:8 Diagram showing the plan of a sewage works (not to scale).

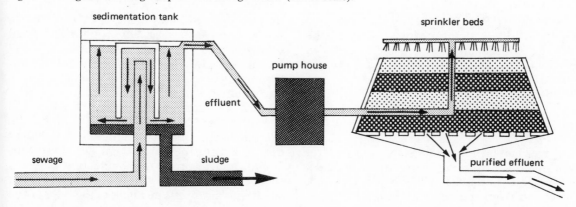

Fig. 19:9 A modern sewage works showing a sprinkler bed.

To avoid these problems, raw sewage can be treated beforehand so that the final products are harmless. One method often used in towns and cities is to utilise the action of bacteria under controlled conditions at a sewage works. The raw sewage from houses passes down concrete sewers to the sewage works (Fig. 19:8). Solid matter is first ground up, screened and then passed into a sedimentation tank from which the solids which form a sludge at the bottom are removed every few days. This can be processed and used as a valuable fertiliser. The effluent from the tank is then pumped to sprinkler beds where rotating arms spray it on to beds of stones or clinker through which the fluid percolates. Bacteria form a film over the clinker and as plenty of oxygen is present, they break down the organic matter very quickly into simpler substances which are safe to pass into the outlet to a river or the sea.

Examine a drop of 'activated' sludge from a sewage works under the high power of the microscope. What types of bacteria can you recognise? What other organisms are present besides bacteria? Why do you think they are present?

Man's use of bacteria

In the dairying industry bacteria are used for a number of purposes. When cheese is made, bacteria are necessary for the formation of lactic acid from the sugar in milk; the lactic acid causes the milk solids to coagulate. In the formation of butter, the cream from which it is made is inoculated with certain bacterial strains which act on various substances present and so give the butter its characteristic flavour. Different strains of bacteria are also used in the making of yoghurt.

Bacteria are also essential for the formation of **silage**, an excellent fodder for dairy cattle. Silage is made from hay which is cut green and put in a silo. Bacteria ferment the sugar in the plants to form lactic acid and anaerobic conditions are produced which prevent the bacteria which cause decay from acting, hence the product retains much of the food value of the original plants and may be used in the winter when fresh grass is not available.

In the manufacture of vinegar, fruit juice is first fermented by yeast and then acetic acid bacteria are used to change the alcohol present into acetic acid. It is the latter that gives the sharp taste to vinegar.

Bacteria are also used for separating the fibres from the stems of the flax plants before they are processed as linen threads, an operation known as **retting**. In the leather industry bacteria are also used during the tanning of hides and in the tobacco industry in the curing of tobacco leaves; in the latter process the fermenting action of the bacteria produces special flavours.

BACTERIA AS PARASITES

Bacteria and the spread of disease

So far we have been mainly concerned with the bacteria that can cause decay; now we come to those which cause disease.

Louis Pasteur, apart from his important work on decay, did much to increase our understanding of how bacteria cause diseases. His work on anthrax and hydrophobia (rabies) is thrilling to read about. However, it was a German doctor, **Robert Koch** (1843–1910), who first proved conclusively that a particular bacterium caused a certain disease. He also worked on anthrax, a disease which killed great numbers of domestic animals. When he examined the organs of animals which had died of anthrax he found them swarming with bacilli, and to prove the connection between them and the disease, he transferred some of them into a tiny cut he made at the base of the

Fig. 19:10 Robert Koch (1843–1910).

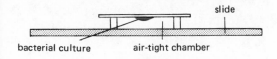

Fig. 19:11 Diagram illustrating the principle of the 'hanging drop' method of culturing bacteria.

tail of a healthy mouse. The mouse soon contracted anthrax and died. But he wanted to know what happened to the bacilli inside the mouse; how did they multiply so fast? Could he possibly grow them outside the body of an animal and watch them under the microscope? He thought of an ingenious method of doing this; he would try to grow them in a drop of colourless fluid from the eye of a freshly-killed ox as this was a nutrient fluid of animal origin. To observe the bacteria, he thought out a technique which is still practised today known as the 'hanging drop' method (Fig. 19:11). Under the microscope, after hours of watching, he saw the minute rods dividing and growing, dividing and growing. Then, when the drop was crowded, he transferred a few to another drop, using a sterilised splinter of wood, and in this way found that he could sub-culture them time and time again. He now had cultures which had never been in the body of any animal. What would happen if he put some into a cut in another healthy mouse? He did this and the mouse died of anthrax! He had at last proved that this particular type of bacillus was specific in its action.

When you grew bacteria on agar plates, you probably discovered that bacteria were present on the skin and in your breath when you coughed; if you tested material from your teeth or finger nails you will have found bacteria there too. Bacteria are also present in our nose and throat and in the gut. Most bacteria are harmless, some in the gut are very useful (p. 186), but a few may cause disease. A disease which spreads from person to person is said to be **infectious**. How are bacteria spread? By studying the various methods of infection we can deduce sensible ways of reducing their effect.

1. Airborne infection

This occurs in droplets of saliva. When a person coughs or sneezes minute droplets are sprayed into the air which may then be breathed in by others. Epidemics spread quickly in this way, especially in crowded places such as classrooms, cinemas and public transport. Good ventilation reduces the risk of infection. In the case of dangerous infectious diseases which are transferred in this way the patient is completely isolated to prevent the spread of the disease.

2. Contact infection

Diseases spread by direct or indirect contact with the body of an infected person are said to be **contagious**. They may be spread when such things as towels, hair brushes and clothes are shared by others. Impetigo, a very infectious disease of the skin which often occurs in schools, is quickly spread in this way. Not all contagious diseases are due to bacteria; some, such as ringworm and athlete's foot, are caused by fungi.

3. Food infection

Some conditions often described as 'food poisoning' are caused by contamination of food by bacterial spores. This may occur if food is handled by somebody with dirty hands in food shops, restaurants or kitchens. Diseases such as typhoid can be spread in this way by a **carrier**, i.e. somebody who harbours the bacteria but does not have the symptoms of the disease. Stored food can also become contaminated by rats or mice or, when displayed in shops, by flies. The packaging of food has done much to lessen some of these risks. Shops and factories in many countries are regularly inspected to see that legal standards of cleanliness are maintained.

4. Infection through drinking water

Cholera, typhoid and paratyphoid may be spread if drinking water becomes contaminated by sewage, as the organisms concerned occur in the faeces of infected persons. These diseases are a great scourge, especially in the tropics and in countries where standards of hygiene are low. Here, epidemics of cholera and typhoid may kill vast numbers of people. They also become a major hazard after disasters such as floods and earthquakes when the drinking water is liable to become contaminated. It should be noted that these two diseases may be air-borne and food-borne, as well as being transmitted through drinking water.

Every year more and more water is needed in towns and cities for domestic and industrial use and often it is in short supply. It may be drawn from lakes, reservoirs or rivers and must be purified before it is used for drinking. In some large cities the same water is used many times before it finally passes into the river and so to the sea; each time it has to undergo treatment.

Bacteriologists employed by water boards test mains water at frequent intervals to see that no harmful bacteria are present in sufficient numbers to cause disease; chlorine, which can sometimes be tasted in tap water, is added in sufficient quantities to destroy them.

Disinfectants

The use of chlorine in the water supply is an example of the use of chemicals to destroy micro-organisms. Such substances are called **disinfectants**. A Scottish doctor, **Joseph Lister** (1827–1912), was the first to use a disinfectant to kill bacteria during a surgical operation. Surgery in those days was extremely dangerous, because even if the patient survived the shock of the operation, there was a great likelihood of death soon after, due to the

Fig. 19:12 Joseph Lister (1827–1912).

wounds becoming septic, i.e. **gangrenous**. Lister believed that sepsis was due to bacteria in the air getting into the wound and causing the tissues to decay just as the broth had done in Pasteur's flasks. In 1865 he put his theory to the test by carrying out his operations under a fine spray of carbolic, a very strong disinfectant. The results showed that he was right, the wounds did not become septic, but the carbolic was so strong that it also destroyed the tissues, and healing was slow. After further experiments using milder solutions it became accepted practice to use what were then called **antiseptics** in surgery.

Today, types of antiseptics are used which are much more efficient, as they do little damage to living tissues. When the skin is broken by cuts or grazes it is always wise, after washing the area, to apply one of these modern disinfectants and cover the wound to prevent further invasion of bacteria from the air.

In surgery today antiseptics are still used, but to a much more limited extent as now the

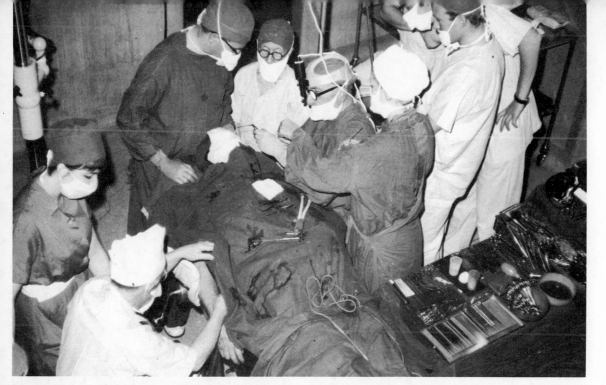

Fig. 19:13 A surgical operation in progress.

emphasis is on **asepsis**—surgery carried out under bacteria-free conditions.

Surgical operations

The operating theatre is designed to exclude bacteria. The walls and floors are smooth and non-absorbent, so that they can be washed easily with disinfectant; the corners are rounded so that no dust can collect and there are no shelves for the same reason. The fittings are mainly chromium-plated or glass, and easily cleaned. The air supply is filtered so that no dust can enter and the temperature is regulated to prevent shock to the patient. The instruments and all the dressings are all heat sterilised under pressure beforehand.

There is still the risk of contamination by the surgeons, anaesthetists, nurses and the patient; they may all bring germs in with them. To minimise this risk hands and arms are washed thoroughly, sterilised gowns and gloves are put on and a gauze mask is used to cover the nose and mouth. The patient is also prepared for the operation, the area of skin around the point of incision being shaved of hair and treated with a mild antiseptic, and a minimum of sterilised clothing is worn. By taking these extreme precautions against bacterial infection, surgery has largely become free from the dangers of sepsis.

Bacteria and our health

Much is done by local authorities and others to provide us with air, food and water which is free from harmful micro-organisms, but we still catch a lot of diseases which could be avoided if we were more careful over personal hygiene. Unfortunately we are not born with a set of instincts to keep the necessary health rules, so we need to acquire good habits of hygiene which often seem a bore until the reasons for them are understood. You should now be able to see the reasons for the following simple health rules:

1. Hands and nails should be washed before eating and preparing food.
2. Hands should be washed after using the toilet.
3. Hair should be washed and brushed regularly.
4. Baths or showers should be taken after games or other strenuous exercise.
5. Clothes, especially underclothes and socks, should be changed frequently.
6. Teeth should be cleaned regularly.
7. Brush, comb, towel and tooth-brush should not be borrowed or shared.
8. Food should be kept covered and pests such as flies, cockroaches, rats and mice should be destroyed.

9. Disinfectants should be used in lavatories and drains.

No doubt you will be able to think of some more.

Immunity against disease

When there is an epidemic, not everybody catches the disease. It is not just a matter of chance who gets it, because some people have more resistance to infection than others. Resistance to disease is a complex subject as many factors are involved, but one factor of particular importance is **immunity**.

Immunity is resistance to a particular infection; it may be present at birth or acquired during life by natural or artificial means.

1. Natural immunity

This occurs after we have recovered from some infection of bacterial or virus origin. If, for example, somebody becomes ill with diphtheria, the bacteria concerned produce poisonous substances called **toxins** which cause the symptoms of the disease. White blood corpuscles in the blood respond to these poisons by producing **antitoxins** (antibodies) which neutralise the effect of the toxins and so help recovery. These antitoxins prevent any recurrence of the disease as long as they remain in the blood. The duration of the immunity varies considerably according to the disease; it may be for life, e.g. measles, but it is more often for a much shorter period.

2. Artificial immunity

It is probable that during the 18th century 60 million people in various parts of the world died of smallpox, a disease caused by a virus (p. 201), but towards the end of the century an English doctor, **Edward Jenner** (1749–1823), discovered how immunity from the disease could be acquired artificially. During a terrible epidemic in England, Jenner noticed that people such as farmers and milkmaids who had previously contracted **cowpox** did not catch smallpox, while others all around were dying in hundreds. Cowpox is a mild virus disease caught from cattle which causes spots full of pus which are rather similar to those caused by smallpox. Jenner wondered what would happen if he put some of the pus from a cowpox spot into a scratch made on the

Fig. 19:14 Edward Jenner (1749–1823).

arm of a boy who was healthy. He tried this experiment and the boy caught cowpox! (This happened long before Pasteur's experiments.) Then came the crucial question: having had cowpox would the boy contract smallpox or not? There was only one way to find out; he would have to inoculate the boy in a similar way with pus from a patient with smallpox. It was a grave risk, but he did it. The next few days were an anxious time for all concerned, but all went well; the boy did not get smallpox! To make doubly sure Jenner repeated the inoculation several months later, but again there was no effect. We now call this process **vaccination.** Today, owing to the recent practice of vaccinating children when a few months old, this dangerous disease has been largely eliminated. However, in a few countries there are still spasmodic outbreaks, so in order to reduce the risk of an epidemic most countries now make it illegal for travellers to enter unless they have been vaccinated within the previous three years.

Nearly a hundred years later, Pasteur made rather similar experiments with anthrax, having probably got the idea from Jenner's work. In order to study the disease he cultured the anthrax in flasks of chicken broth. He found that if a little of this was injected into an animal it quickly died, but quite by accident he found that if a stale culture, 3 or 4 weeks old, was used the animals contracted some of the symptoms, but quickly recovered. Pasteur did many other experiments before he thought of injecting fresh culture into those animals which had previously been inoculated with the stale culture, using others which had not been so treated as a control. The experiment was a brilliant success; those animals which had not been previously treated all died, while the others kept perfectly healthy.

These experiments showed that immunity from both smallpox and anthrax was brought about by using a mild strain or a weakened form of the organism concerned. However, it has not been found possible to produce weakened forms of many other infections.

Today, two additional techniques are used: a) A pure culture of the organism is grown and the toxin is separated from the living agent by filtration, and a carefully controlled toxin is inoculated into a person to cause the body to produce its own antitoxin. Sometimes, as in diphtheria, the toxin has to be reduced in potency by various treatments before it is inoculated.

When we are inoculated against some diseases a reaction is set up in the body, for example, the arm may become swollen and sore, or a mild fever may develop for a few hours.

b) The toxin is periodically inoculated into an animal such as a horse (why not a smaller animal?) which builds up large quantities of antitoxin in its blood. Blood serum from the horse can then be inoculated into a person.

Both methods are used for diphtheria; the first is the best for producing immunity while the second is used to counteract the action of the toxins in a person who actually has the disease.

Inoculation against diphtheria on a massive scale has reduced the danger of this disease enormously. In the 1930's nearly 3,000 children died of it every year in England alone, now it is extremely rare.

Chemotherapy

Since time immemorial, chemicals, often extracts from plants, have been used in the treatment of disease. Some were effective, others were not. However, most were concerned with the symptoms; they did not kill off the organisms themselves. But as soon as it was realised that many diseases were caused by micro-organisms the search started for chemicals which would kill the organisms, but not the cells of the body at the same time.

Paul Ehrlich (1854–1914) was the first person to succeed. After years of research on many chemical substances, he produced one called **salvarsan** which was the 606th substance he had tried! It was used in the treatment of syphilis. Since then (1910) a number of substances have been discovered and tested and some are still used most effectively today, but many have now been replaced by **antibiotics**.

Antibiotics

These are chemical substances produced naturally by certain fungi and micro-organisms. When you grew bacteria on agar you probably grew some fungi as well. **Alexander Fleming** (1881–1955) did very much the same thing in 1929 when he was culturing staphylococci on agar. He noticed that one culture had somehow become contaminated by a mould fungus. This accident was of great significance as he noticed that there was a clear area around the mould where no bacteria were growing and he jumped to the correct conclusion that the mould was secreting a substance that prevented the growth of the bacteria. The mould was identified as *Peni-*

Fig. 19:15 Alexander Fleming (1881–1955).

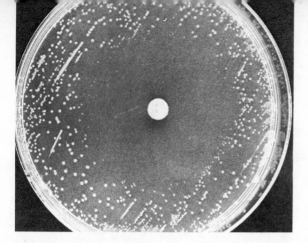

Fig. 19:16 Photograph of a plate of agar inoculated with bacteria showing the effect of penicillin on the bacterial colonies.

cillium notatum, a fungus allied to the green mould you often see on oranges, and its secretion was called **penicillin**, the first of the antibiotics.

It was not until 1938 that **H. W. Florey** and **E. B. Chain** at Oxford continued the research on penicillin to investigate its possible use in medicine for destroying pathogenic bacteria. They knew that penicillin had great potentialities, so when the war came in 1939 they made an all-out effort to produce it on a large scale. To speed up the research Florey flew to America and persuaded a team of scientists in Illinois to collaborate in the project. Eventually a method was found to produce pure penicillin on a large scale and this 'miracle drug' was made available in time to save vast numbers of lives before the war ended.

Since then other antibiotics have been discovered, an important one being **streptomycin**. It has been very effective in reducing tuberculosis in the developed countries from a major cause of death to a disease of minor significance.

Unfortunately it has been found that some bacteria after many generations may no longer be affected by certain antibiotics, i.e. they become **resistant** to it. It is therefore sometimes necessary when treating patients to find out which antibiotic is best to use. When, for example, a person is very ill with an infection of the lungs the doctor takes a sample of the sputum on a sterilised swab, places it in a sterilised tube and sends it to the pathology laboratory of a hospital. Here, agar plates are inoculated from the swab and incubated, and the resulting colonies of bacteria are tested with various antibiotics to see which kills

them. The patient's doctor is informed and the correct treatment can then be given.

You can carry out a similar sort of investigation yourself using penicillin and streptomycin to see which is the more effective in destroying bacteria. Two species can be used, *Staphylococcus albus* and *Escherichia coli*.

For this experiment you will need two sterilised plates of nutrient agar, two sterilised cotton buds, a suspension of *S. albus* and another of *E. coli* and 8 discs which have been impregnated with the antibiotics (see details below).

Cover the surface of the first plate as evenly as possible with the suspension of *S. albus*. Do this by dipping the sterilised cotton bud in the suspension, and using the minimum amount of fluid, stroke the bud over the surface of the nutrient agar to form a series of loops (Fig. 19:17). In the same way, inoculate the second plate with the suspension of *E. coli*, using the second cotton bud. Label each plate.

To each plate add the following four discs, using sterilised forceps:
1. Penicillin of 5 unit strength
2. Penicillin of 10 unit strength
3. Streptomycin of 10 µg strength
4. Streptomycin of 25 µg strength
Fix the lid on each by putting 'Sellotape' round it and incubate the plates for two days at 37°C. A clear region round a disc will indicate the absence of bacteria and hence the effectiveness of the antibiotic.

What conclusions can you draw concerning 1) the action of each antibiotic on the two species of bacterium, 2) the effect of differences in concentration of the antibiotics?

It has recently been discovered that one in five of us has a type of bacterium growing on our skins which has remarkable properties. If you grow a culture of this species and soak up a minute amount of it on a disc of sterilised filter paper and transfer it, as you did with the penicillin disc, to a plate of *Staphylococcus*, a strange thing happens. When the plate is incubated and the colonies grow a clear patch forms round the skin colony. These bacteria are producing their own antibiotic! You could describe it as chemical warfare between different species of bacteria.

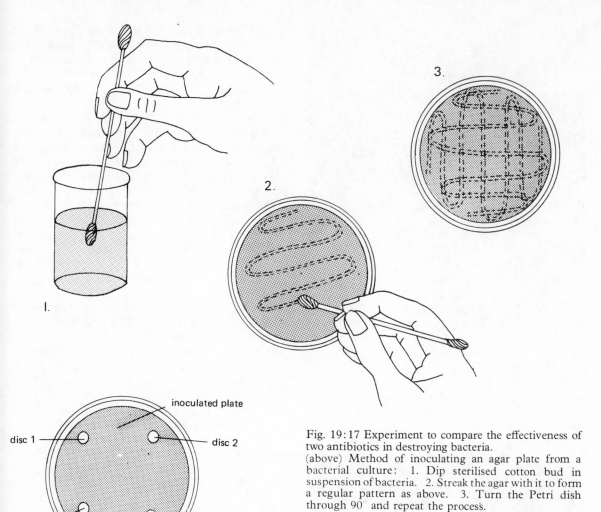

inoculated plate

disc 1

disc 2

disc 3

disc 4

Fig. 19:17 Experiment to compare the effectiveness of two antibiotics in destroying bacteria.
(above) Method of inoculating an agar plate from a bacterial culture: 1. Dip sterilised cotton bud in suspension of bacteria. 2. Streak the agar with it to form a regular pattern as above. 3. Turn the Petri dish through 90° and repeat the process.
(left) The antibiotic-impregnated discs in place.

It has been found that hospital patients who have this species on their skin are far less likely to have an infection in their wounds following an operation than those who have not. Fortunately, this useful species can be transferred from one person's skin to that of another so it is possible to pass on this method of protection. So, some of us carry our antibiotic factory with us wherever we go!

VIRUSES

Viruses are incredibly small, far smaller than the smallest bacteria. The largest can just be seen under the best light microscopes, but the majority have never been seen directly, only photographed under the electron-microscope. Some are rod-shaped, others are many-sided crystals, often with a tail-like projection. The smaller viruses act like chemicals and can be crystallised. Viruses differ from typical living cells in having no nucleus, cytoplasm or surrounding membrane, the smaller ones at least consisting merely of a protein envelope enclosing a giant **nucleic acid** molecule. Nucleic acids occur in viruses in two distinct forms known as DNA and RNA. They are similar to the genetic material found in all plant and animal cells.

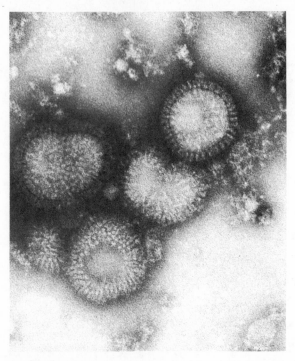

Fig. 19:18 Electronmicrograph of the influenza virus. × 200,000.

Fig. 19:19 Electronmicrograph of the virus causing tobacco mosaic. × 192,000.

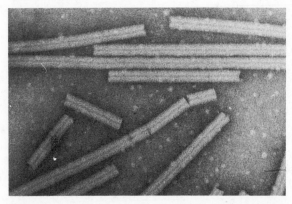

Most viruses cause disease in plants and animals and many of the common infections of man are caused by them. We have already referred to smallpox (p. 198) and yellow fever (p. 73); others include influenza, the common cold, poliomyelitis, measles, German measles, chickenpox, glandular fever and mumps.

Viruses are inactive unless they penetrate living cells. Some invade the cells of bacteria, others plant cells and others, human and animal cells. They are often very specific in their action; in man, for example, one kind of virus will attack only cells in the skin (producing warts), another only certain nerve cells (polio) and another the cells of the salivary glands (mumps).

As viruses have no cytoplasm they produce no enzymes and hence are unable to carry out such functions as respiration, which are associated with living things. However, they have the power of multiplication. This multiplication is not the same as normal reproduction, as the virus does not grow or divide and cannot reproduce on its own, but only with the help of living protoplasm from a plant or animal cell. The way this happens has been discovered in viruses which attack bacteria; these are often called **bacteriophages** or just **phages**.

Phages become attached to the bacillus by their 'tail' (Fig. 19:20). The bacterial membrane dissolves and the DNA content of the phage passes into the cell leaving the protein coat behind. Inside the bacillus the virus DNA makes replicas of itself, using for the purpose the nucleic acids of the bacterium and other substances in the cell. New protein envelopes are then formed, the cell bursts and about 200 exact copies of the phage come out and infect other cells. Each cycle takes about 45 minutes. In this way the nucleic acids of the bacterial cells which are normally used for their own reproduction are 'commandeered' to make copies of the phages instead. In a similar way, if we catch influenza, the virus attacks *our* cells and makes use of the nucleic acids in these cells to produce more virus particles. These attack more cells and the cycle is repeated until vast numbers of cells are infected. Influenza is also extremely infectious and spreads on droplets of saliva from person to person. This is why it is important, for the sake of others, to stay at home if you catch the disease. You will know how influenza periodically spreads right across the world as a serious epidemic. It is estimated that during the epidemic of 1918–19 which followed the First World War half the population of the world caught influenza and over 20 million people died as a result. But influenza, like many virus diseases, is very variable; sometimes it is very virulent, at other times much milder. This may be because there are dif-

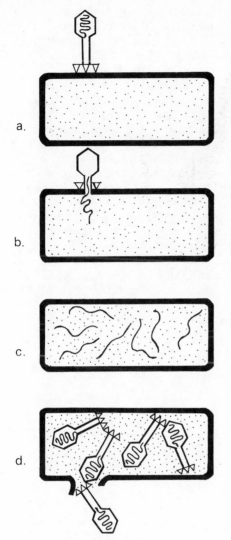

Fig. 19:20 Multiplication of a bacteriophage:
a) Attachment of phage to wall of bacterium.
b) DNA of phage passing into bacterium.
c) Replication of phage DNA.
d) Formation and escape of new phages.

ferent strains of the virus, known as influenza A and B, and variations in a strain may come about by a mutation, a change in the composition of the DNA core of the virus.

Artificial immunity against virus diseases

We saw in the case of smallpox how a mild form occurred in cows called cowpox and immunity was produced by vaccination with cowpox virus. Much research has been done to find suitable vaccines for other virus infections, but one of the difficulties has been to grow the virus artificially. You cannot grow a virus on agar as it only multiplies within living cells, but it has been found possible to grow animal cells outside the body in bacteria-free nutrient material and to infect these cultures with a virus. An American scientist, **Dr Jonas Salk**, used this technique when working on **poliomyelitis**. He 'grew' the virus in a tissue culture of cells from a monkey's kidney and then inactivated the virus with formalin. This prevented further multiplication, but the factor was retained which caused the white blood cells of the person vaccinated to form its own antibodies. This vaccine gave a high degree of immunity.

Other treatments of virus diseases

Treatment of viruses by antibiotics is of use only with some of the largest viruses, such as the one causing **trachoma**, a very common cause of blindness in hot countries. What scientists are looking for is something which will prevent the virus from replicating its nucleic acid. A significant break-through occurred when it was found that when two different viruses invade the same cell one somehow interfered with the multiplication of the other. In 1957 **Isaacs** and **Lindemann** discovered that when this happened a substance was made which they were able to isolate, which they called **interferon**. It does not directly destroy the other virus or prevent it from entering the cell, but it does interfere with its multiplication. Its great advantage is that it is naturally produced, is non-poisonous to body cells and affects a wide range of viruses. Research continues on this very exciting substance.

Plant viruses

The first virus ever to be isolated was the cause of tobacco leaf mosaic, a disease of the tobacco plant in which the leaves have a mottled pattern of pale green and yellow where the cells have been attacked. In 1935 **Dr Wendell Stanley** made an extract from the diseased tobacco leaves and from it obtained crystals of the virus, which he then injected into healthy plants causing them to develop the disease. This virus is so stable that it can survive for many years in cured tobacco and people who handle it may spread the virus to other crops such as tomatoes which are susceptible to it.

You may have seen tulip flowers with attractive streaks of a different colour in the petals. This variegation is due to a virus attacking some of the cells and is of interest because it is shown in Dutch paintings going back to the 16th century, and thus represents the oldest plant virus known.

Many plant viruses are spread from plant to plant by aphids and their allies. When they feed on the sap of plants they plunge their mouthparts into the tissues and suck up the juices, and on going to another may transfer the virus. Plants attacked by a virus are usually weakened in consequence.

The study of viruses (virology) is a very exciting and fast-growing section of biology which has many implications for the future welfare of man. The connection between viruses and certain forms of cancer is currently being investigated, and also the production and application of new vaccines to combat other diseases. Research is also being carried out on the use of viruses in the control of insects and other pests.

Appendix

The list of chemical recipes refers only to those chemicals and reagents for which insufficient information is given in the text. The books which have been selected are suitable for pupils, either for general reading or for reference. The list of 8 mm loop films and 16 mm films includes a small selection of those considered to be particularly relevant to this course, but there are many others which would also be useful. All the apparatus mentioned in the text can be obtained from leading biological suppliers.

For teachers, much useful information relating to experimental techniques, bibliographies etc. can be found in the Nuffield Biology Teachers' Guides (Longmans/Penguin) and the Teachers' Guides to the Scottish Course, Biology by Inquiry (Heinemann). The Laboratory Book (Nuffield Advanced Biological Science, Ed. Peter Fry, Penguin Books 1971), includes details of how to culture living organisms, chemical recipes, sources of apparatus and living organisms etc. which are relevant to 'O' level work.

Chemical Recipes

Ascorbic acid (vitamin C)
Can be obtained in tablets each containing 50 mg. One tablet dissolved in 50 cm^3 of distilled water will give an ascorbic acid solution of approximately 0·1%. Crush the tablets in a mortar and filter the solution before use.

Benedict's solution
Solution A: Dissolve 173 g of hydrated sodium citrate and 100 g of hydrated sodium carbonate in approximately 800 cm^3 of warm distilled water. Filter through glass wool into a measuring cylinder and make up to 850 cm^3. Solution B: Dissolve 17·3 g of hydrated cupric sulphate in 100 cm^3 of cold distilled water. Pour solution A into a large beaker and add solution B slowly, stirring continuously with a glass rod. Make up the total volume to 1 litre.

Bicarbonate indicator solution
A detailed discussion may be found in the Nuffield Biology Teachers' Guide III, Chapter 1. A stock solution should be prepared as follows: Dissolve 0·2 g of thymol blue and 0·1 g of cresol red in 20 cm^3 of ethanol. Then dissolve 0·84 g of sodium bicarbonate (Analar) in 900 cm^3 of distilled water in a large beaker. Add the cresol red/thymol blue solution to the bicarbonate solution, transfer it to a 1 litre volumetric flask and add distilled water to make up exactly 1 litre.

Prepare a working solution by pipetting exactly 25 cm^3 of the stock solution into a 250 cm^3 volumetric flask and adding distilled water up to the 250 cm^3 mark. Before use the solution should be aspirated with fresh atmospheric air so that it is in equilibrium with the carbon dioxide in the atmosphere. It should be a deep red colour when seen in a flask and orange-red when in small quantities in a test tube.

DCPIP solution (dichlorophenol-indophenol)
A 0·1% aqueous solution is suitable for determining the presence of vitamin C in foods.

Iodine/potassium iodide solution
Dissolve 6 g of potassium iodide in about 200 cm^3 of distilled water. Add 3 g of iodine crystals and make up to 1 litre with distilled water.

Millon's reagent
Solution A: Very carefully pour 100 cm^3 of pure concentrated sulphuric acid into about 800 cm^3 of distilled water in a flask. Cool under a tap. Grind 100 g of mercuric sulphate in a mortar and dissolve it in the diluted acid. Filter and make the solution up to 1 litre with distilled water. Solution B: Prepare a 1% aqueous solution of sodium nitrite. (This is slightly unstable.) The working solution is most easily prepared by adding 2 volumes of solution A to 1 volume of solution B. It will only remain stable for a few weeks. Note: This reagent is poisonous.

Potassium hydroxide solution (for gas analysis experiments)

Dissolve 100 g of potassium hydroxide pellets in 50 cm^3 of water to make a saturated solution. As a considerable amount of heat is produced, cool the solution under a tap.

Potassium pyrogallate (alkaline pyrogallol)

Small quantities can be made up easily by placing 1 volume of potassium hydroxide pellets and 5 volumes of solid resublimed pyrogallol in a beaker. Put 2 volumes of water and 1 volume of liquid paraffin into another beaker, pour it quickly on to the solids, and stir gently with a glass rod. The liquid paraffin will float and prevent atmospheric air from entering. The aqueous solution of potassium pyrogallate below should be nearly colourless.

Yeast/glucose solution (for experiments on anaerobic respiration)

A 5% solution of yeast in a 10% solution of glucose is suitable.

Additional notes for teachers

Chapter 12, p. 122. Visking tubing 14 mm in diameter is suitable.

Chapter 14, p. 140. J tubes (capillary tubes for gas analysis) are obtainable from suppliers such as T. Gerrard and Co. Ltd., Griffin and George Ltd. and Philip Harris Biological Ltd.

Chapter 19, p. 200. Antibiotic discs may be obtained from Oxoid Division, Oxoid Ltd., Southwark Bridge Road, London, SE1 9HF. See also Nuffield Biology Guide II for a detailed discussion of microbiological methods.

Books

Keys for identification may be found in:

Nuffield Science Teaching Project (1966): Keys to Small Organisms in Soil, Litter and Water Troughs. Longmans/Penguin Books.

Reid, D. and Booth, P. (1970). Biology for the Individual: 1 Sorting Animals and Plants into Groups. Heinemann.

Sankey, J. (1958). A Guide to Field Biology. Longmans.

General identification books:

Blandford Colour Series: e.g. Birds, Insects, Mushrooms and Toadstools, Pond and Stream Life, Seashore Life and others. Blandford Press.

Collins Pocket Guides: e.g. Pocket Guide to British Birds, the Seashore, Wild Flowers; Field Guide to the Birds of Britain and Europe, Field Guide to the Butterflies of Britain and Europe. Collins.

The Observer Series: e.g. Birds, Butterflies, Common Insects and Spiders, Larger Moths, Pond Life, Trees. Warne.

The Oxford Books: e.g. Flowerless Plants, Insects, Invertebrates, Vertebrates. Oxford University Press.

The Young Specialist Series: e.g. Birds, Butterflies, Mammals, Pond Life, Trees, Wild Flowers. Burke.

The following reference books may also be useful:

Macan, T. (1959). A Guide to Freshwater Invertebrate Animals. Longmans.

Martin, W. Keble (1965). The Concise British Flora in Colour. Ebury Press and Michael Joseph.

Mellanby, H. (1963). Animal Life in Fresh Water. Methuen.

Newman, L. Hugh, and Mansell, E. (1968). The Complete British Butterflies in Colour. Ebury Press and Michael Joseph.

For background reading:

Aykroyd, W. B. (1964). Food and Man. Pergamon.

Barnett, A. (1961). The Human Species. Penguin.

Brown, E. S. (1955). Life in Fresh Water. Oxford University Press.

Butler, C. G. (1962, revised). The World of the Honeybee. Collins.

Clegg, J. (1965 3rd edition). The Freshwater Life of the British Isles. Warne.

Dale, A. (1951). Patterns of Life. Heinemann.

De Kruif, P. (1930). Microbe Hunters. Cape.

Devon Trust for Nature Conservation (1972). School Projects in Natural History. Heinemann.

Dodd, E. F. (1956). The Story of Sir Ronald Ross and his Fight against Malaria. Macmillan.

Ford, E. B. (1972, revised). Butterflies. Collins.

Gray, J. (1959). How Animals Move. Penguin.

Harrison, R. J. (1958). Man the Peculiar Animal. Penguin.

Howard, E. (1948). Territory in Bird Life. Collins.

Imms, A. D. (1971, revised). Insect Natural History. Collins.

Knight, M. (1960). The Young Field Naturalist's Guide. Bell.

Lack, D. (1946). Life of the Robin. Witherby. (Also published as a Collins Fontana paperback.)

Lawrence, M. and Brown, R. (1973, revised). Mammals of Britain: Their Tracks, Trails and Signs. Blandford.

Leutscher, A. (1963). A Study of Reptiles and Amphibians. Blandford.

Matthews, L. H. (1952). British Mammals. Collins.

Neal, E. (1948). The Badger. Collins.

Prime, C. T. (1963). The Young Botanist. Nelson.

Rook, Arthur (Ed.) (1964). The Origins and Growth of Biology. Penguin.

Smith, Malcolm (1954, revised). The British Amphibians and Reptiles.

Southwood, T. R. E. (1963). Life of the Wayside and Woodland. Warne.

Taunton, J. (1968). Bird Projects for Schools. Evans Brothers.

Walker, K. (1956, revised). Human Physiology. Penguin.

Williams, H. (1961). Great Biologists. Bell.

8 mm Loop Films

Many of these titles can be purchased in both standard 8 and super 8 formats. Full details can be obtained from the distributors.

Methods of Finding Animals 1 (Beating and sweeping). Gateway.

Methods of Finding Animals 2 (Sieving leaf litter and insect light trap). Gateway.

Movement in Protozoa. Gateway.

Development of the Locust 1: Hatching and Growth (Nuffield Foundation). Longmans.

Development of the Locust 2: Pairing and Egg-laying (Nuffield Foundation). Longmans.

The Locust as a Biting Insect (Nuffield Foundation). Longmans.

Cabbage White Butterfly: The Egg and Caterpillar. Gateway.

Cabbage White Butterfly: The Chrysalis and Butterfly. Gateway.

The Housefly as a Sucking Insect (Nuffield Foundation). Longmans.

Breathing. Macmillan.

Chewing in Carnivores (Nuffield Foundation). Longmans.

Human Teeth. Gateway (in collaboration with the General Dental Council).

Comparison of Teeth in Mammals. Ditto.

Structure of Teeth. Ditto.

Dental Decay. Ditto.

Swallowing. Macmillan.

Rat Dissection 1: Alimentary Canal. Gateway.

Sewage Organisms. BBC Publications.

16 mm Films

Angiosperms: the flowering plants. Encyclopaedia Britannica, distributed by Rank.

The Ruthless One. (The life cycle and biology of the locust.) P.F.B.

Life History of the Cabbage Butterfly. E. R. Skinner and G. H. Thompson, distributed by E.F.V.A.

Social Insects: the honey bee. Encyclopaedia Britannica, distributed by Rank.

Dance of the Bee. B.F.I.

Fish. Encyclopaedia Britannica, distributed by Rank.

What is an Amphibian? Ditto.

Swallows. Royal Society for the Protection of Birds, distributed by Gateway.

The Kingfisher. BBC, re-edited and distributed by Gateway.

What is a Bird? Encyclopaedia Britannica, distributed by Rank.

What is a Mammal? Ditto.

Microscopic Life in Soil. Stanton Films, distributed by Gateway.

Addresses:

BBC Publications, 35 Marylebone High Street, London W1M 4AA.

British Film Institute (Film Library), 42 Lower Marsh, London SE1 7RG.

Educational Foundation for Visual Aids, The National Audio Visual Aids Library, Paxton Place, Gipsy Road, London SE27 9SR.

Gateway Productions Ltd., Waverley Road, Yate, Bristol BS17 5RB.

Longman Group Ltd., Pinnacles, Harlow, Essex.

Macmillan and Co. Ltd., Brunel Road, Basingstoke, Hants.

Petroleum Films Bureau, 4 Brook Street, London W1Y 2AY.

Rank Film Library, 1 Aintree Road, Perivale, Greenford, Middlesex.

Acknowledgments

For permission to reproduce photographic illustrations, acknowledgment is due as follows: Figs. 9:5 (Ken Hoy), 11:5b (John Gooders), Ardea Photographics; 19:21, Stewart Bale Ltd; 19:9 (Tony Boxall), Barnaby's Picture Library; 11:9, Blandford Press Ltd., reproduced from *Mammals of Britain: Their Tracks, Trails and Signs*, by M. Lawrence and R. Brown, photo by C. Atherton; 7:17, Dr .C. G. Butler; 11:6c, David Bygott; 1:1 (Jon Blau), Camera Press; 16:10, Canadian High Commission, London; 6:10, 6:11 (C. Ashall), Centre for Overseas Pest Research; 9:4 bottom left (S. C. Bisserot), 9:4 top, 9:4 bottom right, 9:7, 9:9, 9:11a, 11:2d, 11:5c (Jane Burton), 10:13d, 11:6f (Bruce Coleman), 10:14a (Klaus Fiedler), 9:11g (David Hughes), 10:13g (E. Breeze Jones), 10:14c (Gordon Langsbury), 10:13e (John Markham), 11:6b (Graham Pizzey), 11:2e (N. Tomalin) Bruce Coleman Ltd; 11:8, Eva Crawley; 7:19, 11:7, Farmers Weekly; 19:19, Dr John T. Finch; 1:2, Douglas Fisher; 16:8, Food and Agriculture Organisation; 1:4, 2:2a, 2:2b, 2:2c, 2:3 (Dr J. H. Kugler), 2:6, 2:10 (M. I. Walker), 2:17, 4:7, 4:20, 4:21, 4:23, 4:32, 5:2, 5:5, 5:6, 5:7, 5:9, 5:13, 10:8, 13:6, 14:5, 15:1, 15:8, 15:9, 15:10, 15:13, 17:5, 17:9, 17:10, 17:12, 18:9, 18:10, 19:6, 19:16, Philip Harris Biological Ltd; 10:1, 10:13f, Eric Hosking; 7:9, I.C.I. Agricultural Division; 13:7, Keystone Press Agency; 9:10, 11:2c, 11:6d, 11:6e, Geoffrey Kinns; 17:6, Longman, Green and Co., reproduced from *An Atlas of Skiagrams* by Johnson and Symington; 19:3, 19:10, 19:12, 19:14, 19:15, Mansell Collection; 13:8, Monark-Crescent AB; 11:2a, 16:9, Andrew Neal; 16:6, Oxfam; 6:14, 6:16, A. E. McR. Pearce; 15:17, Rothamsted Experimental Station; 4:26, 19:13, St Mary's Hospital Medical School; 6:13, 6:19, 7:5, Shell Photographic Unit; 7:8, 16:5, C. James Webb; 19:2, 19:18, Dr N. G. Wrigley; 10:13h, Zoological Society of London. Also to the following for other illustrations and data: 5:8 (iris rhizome), 7:13, based on Mackean's *Introduction to Biology*, John Murray (Publishers) Ltd; 14:15, from *Smoking and Health Now* (The Royal College of Physicians of London), Pitman Medical and Scientific Publishing Co. Ltd; 15:18, data from Rothamsted Experimental Station Report for 1968 Part 2; 16:4, based on F.A.O. data; 16:6, 16:7, data from Oxfam; tables on p. 141, data from Nuffield Biology Text III, table on p. 157, data from Nuffield Biology Text IV and Teachers' Guide IV, Longman Penguin.

Index

Figures in bold type show that the subject is illustrated, but in addition there may be a further reference in the text on the same page.

212

215